Geology of the Haddington district

The district described in this memoir includes that part of the East Lothian coast from Cockenzie to North Berwick and extends southwards across the gently rising ground to take in the south-west part of the Lammermuir Hills. The district is covered by Sheet 33W and part of Sheet 41 of the geological map of Scotland.

The introductory chapter outlines the physical features of the district and the geological history. The Lower Palaeozoic rocks of the Lammermuir Hills are described, with particular reference to their stratigraphy and structure. Chapters then follow dealing with the sediments of Lower Devonian and Devono-Carboniferous age. Carboniferous sediments underlie the major part of the district and successive chapters give the details of the Lower and Upper Carboniferous sediments, the latter containing a summary of the information on coal-bearing strata gained from past mining activity. A section on the palaeontology of the Carboniferous rocks is followed by details of the igneous rocks in the district which comprise the Garleton Hills Volcanic Rocks of Lower Carboniferous age and the intrusive rocks.

A chapter is devoted to discussing the geological structure of the district and this is followed by a discussion on the Quaternary deposits including their history of deposition and their present distribution. The final chapter is on the economic geology and details the various natural resources present.

Trap featuring in trachyte lavas, modified by ice-action and glacial drainage,
Kae Heughs, Garleton Hills (D 3262) *Frontispiece*

BRITISH GEOLOGICAL SURVEY
Scotland

A. D. McADAM and
W. TULLOCH

Geology of the Haddington district

Memoir for 1:50 000 Sheet 33W and part of Sheet 41

CONTRIBUTORS
R. W. Elliot, J. D. Floyd,
D. K. Graham and R. B. Wilson

1835 Geological Survey of Great Britain
150 Years of Service to the Nation
1985 British Geological Survey

BRITISH GEOLOGICAL SURVEY
Natural Environment Research Council

LONDON: HER MAJESTY'S STATIONERY OFFICE 1985

First published 1985

ISBN 0 11 884452 0

Bibliographical reference

McAdam, A. D., and Tulloch, W. 1985.
Geology of the Haddington district.
Mem. Br. Geol. Surv., Sheet 33W and part of 41, 99pp.

Authors

A. D. McAdam, BSc, MIGeol
British Geological Survey, Murchison House,
West Mains Road, Edinburgh EH9 3LA
W. Tulloch, BSc
Formerly Institute of Geological Sciences, Edinburgh

Contributors

R. W. Elliot, BSc
J. D. Floyd, BSc, PhD
D. K. Graham, BA
R. B. Wilson, DSc, FRSE
British Geological Survey, Murchison House,
West Mains Road, Edinburgh EH9 3LA

Other publications of the Survey dealing with this and adjoining districts

Books

British Regional Geology
The Midland Valley of Scotland, 3rd Edition, 1985

Memoirs
The geology of the Midlothian Coalfield, 1958
The geology of the neighbourhood of Edinburgh, 1962

Geological maps

1:625 000
Geological (North)
Quaternary (North)
Aeromagnetic (Sheet 1)

1:50 000 and one-inch to one mile (1:63 360)
Sheet 25W (Galashiels): Solid 1983
Sheet 32E (Edinburgh): Solid 1977; Drift 1967
Sheet 33W (Haddington): Solid 1983; Drift 1978
Sheet 33E (Dunbar): Solid 1980; Drift 1978

CONTENTS

FIGURES

TABLES

PLATES

PREFACE

The district described in this memoir is covered by Sheet 33W and part of Sheet 41 of the 1:50 000 Geological Map of Scotland. It forms part of the first area to be mapped by the Geological Survey in Scotland. It was surveyed in the late 1850's by H. H. Howell and A. Geikie and the first edition of the map (Sheet 33) appeared in 1860. The first edition of the memoir was published in 1866.

Between 1888 and 1896 B. N. Peach and J. Horne revised the Lower Palaeozoic rocks and from 1902 to 1904 the post-Silurian rocks were resurveyed by E. M. Anderson, E. B. Bailey, G. Barrow, C. B. Crampton and H. B. Maufe. The second editions of the map and the memoir were published in 1910.

Small parts of the district were resurveyed by J. B. Simpson in 1936–38, by Mr N. R. Martin in 1947, by Mr W. Tulloch in 1950 and a full revision was carried out between 1960 and 1970 by Dr M. F. Howells, Mr A. D. McAdam and Mr W. Tulloch, the District Geologists being J. E. Richey, J. B. Simpson, Mr R. A. Eden and Mr G. I. Lumsden.

During the course of the survey, a deep, cored borehole was sunk at Spilmersford which penetrated almost all of the Lower Carboniferous succession and the results of the ensuing investigations were published in 1974 in the Bulletin of the Institute of Geological Sciences, No. 45. Since the publication of the previous edition of the memoir numerous papers have been published on various topics concerning the district and these are referred to in the text. The fossils from field localities were collected by Mr P. J. Brand and Mr D. K. Graham.

The map of the part of the district on Sheet 41 on the 1:63 360 scale was published in 1970. The Drift edition of the map of the whole district was published in 1978 and the Solid edition in 1983 on the 1:50 000 scale.

The present memoir was written mainly by Mr A. D. McAdam and Mr W. Tulloch. Dr J. D. Floyd revised the chapter on the Lower Palaeozoic rocks, Dr R. B. Wilson contributed the chapter on Carboniferous palaeontology, Dr I. Strachan of the University of Birmingham identified the graptolites and Mr D. K. Graham was responsible for the palaeontology of the Quaternary sediments. Mr R. W. Elliot wrote the sections dealing with the petrology of the igneous rocks. Mr D. F. Ball contributed the section on groundwater resources. The photographs were taken by Mr T. S. Bain, Mr A. F. Christie and Mr F. I. MacTaggart. The memoir was edited by Dr R. B. Wilson.

British Geological Survey
Keyworth
Nottingham NG12 5GG

7th August 1985

G. M. Brown
Director

GEOLOGICAL SEQUENCE

The geological formations which occur within the district are summarised below:

SUPERFICIAL DEPOSITS (DRIFT)
Quaternary (Recent and Pleistocene)
Blown sand
Peat
Alluvium and lake deposits
Present beach and saltmarsh deposits
Post-Glacial marine and estuarine deposits
Late-Glacial marine and estuarine deposits
Glacial meltwater deposits
Till

SOLID FORMATIONS		*Generalised thickness* m
Carboniferous		
WESTPHALIAN		
Lower Coal Measures	cyclic sequence of sandstones, mudstones, coals and seatclays	30
NAMURIAN		
Passage Group	sandstones, mudstones, seatclays and thin coals	125
Upper Limestone Group	cyclic sequence of sandstones, mudstones, thin marine limestones, seatclays and thin coals	125
Limestone Coal Group	cyclic sequence of sandstones, mudstones, coals, seatclays and ironstones	160
DINANTIAN		
Lower Limestone Group	cyclic sequence of sandstones, mudstones, marine limestones and thin coals	100
Calciferous Sandstone Measures	upper part predominantly sandstone, with several marine limestones towards the top and a number of marine shell-beds lower down, several thin coals; middle part Garleton Hills Volcanic Rocks, trachytic and basaltic lavas and tuffs; lower part alternating sandstones, mudstones and thin cementstones, reddened below volcanic rocks	600 +
Devono-Carboniferous		
Upper Old Red Sandstone	red, brown and green sandstones, siltstones and mudstones; beds of cornstone in upper part; pebbly and conglomeratic bands in lower part	250 +
Unconformity		
Lower Devonian		
Lower Old Red Sandstone	red-brown conglomerates, bands of pebbly sandstone, thin siltstones and mudstones	30
Unconformity		
Silurian		
Llandovery	greywackes, siltstones, shales and mudstones	not known
Unconformity		
Ordovician		
Caradoc – Ashgill	greywackes with conglomerate bands, siltstones, shales and mudstones, graptolite shales, beds of chert; one thin lava flow, few thin tuffaceous bands	not known
Igneous intrusions		
?Tertiary	tholeiitic olivine-basalt dyke	
Late Westphalian – Stephanian	quartz-dolerite dykes	
Namurian – early Westphalian	olivine-dolerite and teschenite sills	
Dinantian	olivine-dolerite, trachydolerite, basanite and trachyte sills, plugs and dykes; agglomerate, tuff and basanite intrusions in vents	
Lower Devonian	felsite, porphyrite, lamprophyre and granitic dykes and other minor intrusions	

SIX-INCH MAPS

The geological six-inch maps covering, wholly or in part, the 1:50 000 Haddington (33W) Sheet are listed below with the names of the surveyors (M. F. Howells, A. D. McAdam, N. R. Martin, J. B. Simpson and W. Tulloch) and the dates of survey.

The maps are not published but they are available for reference at the British Geological Survey office, Murchison House, Edinburgh where photocopies can be purchased.

NT 35 NE	Middleton	Tulloch	1962
NT 36 NE	Chesterhill	Simpson and Tulloch	1937, 1950
NT 36 SE	Pathhead	Tulloch	1950, 1962
NT 37 SE	Prestonpans	Simpson	1936–38
NT 45 NW	Fala Moor	Tulloch	1962–63, 1967
NT 45 NE	Soutra	Tulloch	1963, 1967–68
NT 46 NW	Pencaitland	Tulloch and Howells	1950, 1961
NT 46 NE	Saltoun	Howells	1961
NT 46 SW	Fala	Tulloch	1962–3
NT 46 SE	Humbie	Tulloch	1963–66
NT 47 NW	Longniddry	Howells	1961
NT 47 NE	Aberlady	Howells	1961
NT 47 SW	Tranent	Howells	1961
NT 47 SE	Gladsmuir	Martin and Howells	1946–47, 1961
NT 48 SE	Gullane	Howells	1961
NT 55 NW	Kelphope	Tulloch	1968–70
NT 55 NE	Hunt Law	Howells	1966–67
NT 56 NW	Gifford	Howells	1962
NT 56 NE	Danskine	Howells	1962–63
NT 56 SW	Lammer Law	Tulloch	1965–69
NT 56 SW	Hopes	Howells and Tulloch	1962–66, 1970
NT 57 NW	Athelstaneford	McAdam	1964–67
NT 57 NE	Markle	McAdam	1964–67
NT 57 SW	Haddington	Howells	1961–62
NT 57 SE	Morham	Howells	1963–64
NT 58 NE	Fidra	Martin and McAdam	1946–47, 1967
NT 58 NE	North Berwick	Martin and McAdam	1946–47, 1967
NT 58 SW	Dirleton	McAdam	1964–65
NT 58 SE	Balgone	McAdam	1965–67

NOTES

In this memoir the word 'district' means the area of land included on the Scottish 1:50 000 Geological Sheet 33W (Haddington) and part of Sheet 41W south of the Firth of Forth.

Numbers in square brackets are National Grid references within 100 km square NT.

Numbers preceded by the letter S refer to the Sliced Rock Collection of the British Geological Survey.

Numbers preceded by the letters D and MNS refer to the British Geological Survey Photograph Collection (Scotland). A full list of geological photographs taken in the district is available on application to British Geological Survey, Murchison House, Edinburgh EH9 3LA.

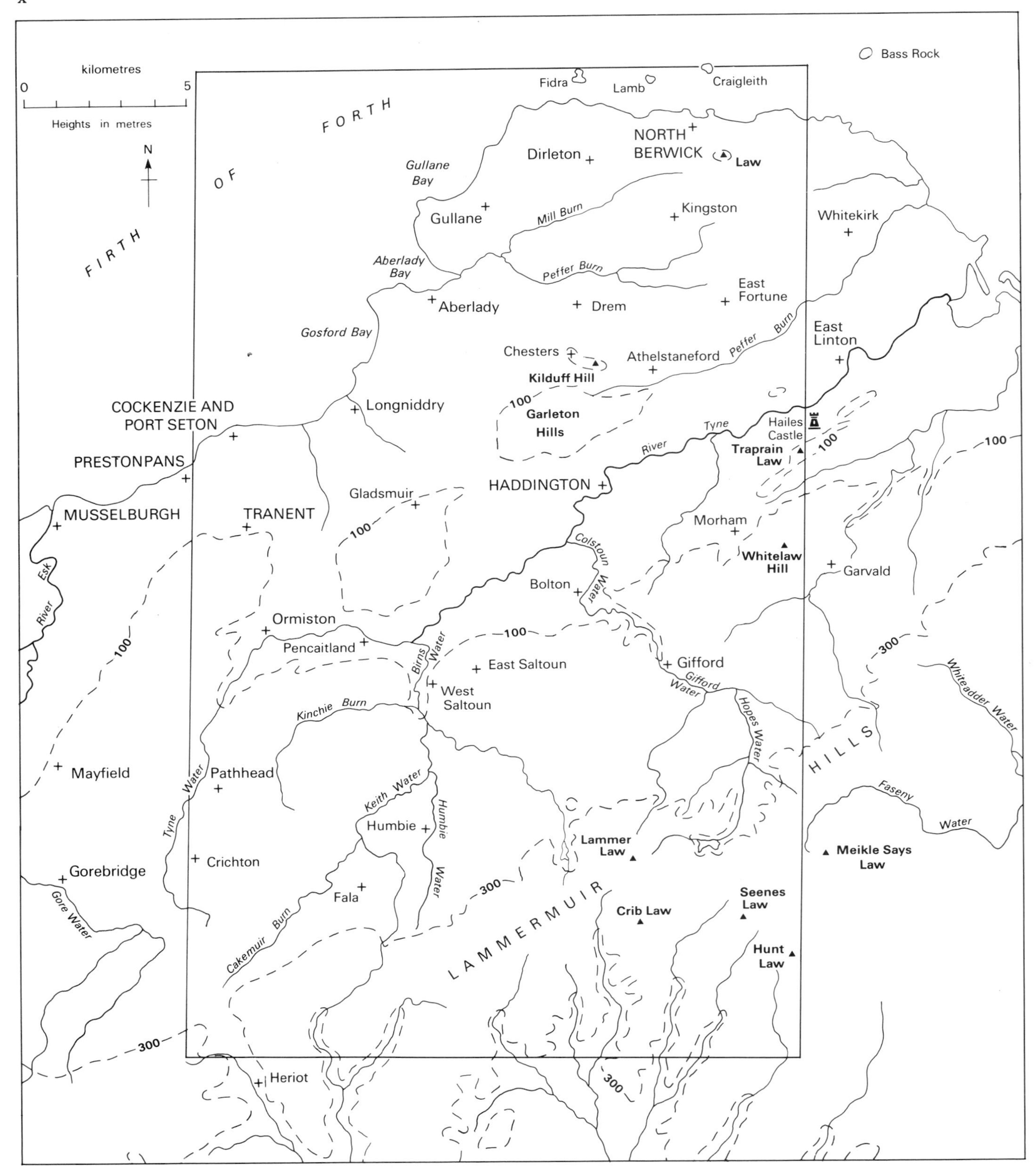

Figure 1 Main physical features of the district and adjoining areas

CHAPTER 1

Introduction

AREA AND PHYSICAL FEATURES

This account describes the geology of the district represented on the Haddington (33W) Sheet along with part of the North Berwick (41) Sheet of the 1:50 000 geological map of Scotland. The district lies south of the Firth of Forth and east of Edinburgh (Figure 1). It falls mainly within the Lothian Region, but in the south a part of the Borders Region is included.

The district slopes gently northwards to the Firth of Forth from the Lammermuir Hills in the south-east, dominated by Lammer Law (527 m), Seenes Law (513 m), Crib Law (509 m) and Hunt Law (495 m). The drainage of the area is mostly into the River Tyne, which flows to the north-east into the Firth of Forth. Almost all its tributaries are from the Lammermuir Hills, the main ones being the Birns Water and the Colstoun Water. The area north of the Garleton Hills, which interrupt the gentle slope to the sea, is drained by two streams, both called Peffer Burn, one flowing east, one flowing west. A major watershed occurs in the south-east corner of the district, drainage to the south of it being into the Tweed basin.

Geologically and geographically the district is divided into two contrasting sections by the Lammermuir Fault (Figure 2), the most easterly fracture in the Southern Upland Fault-belt, which divides the Carboniferous and Devonian sedimentary and igneous rocks of the Midland Valley from the Ordovician and Silurian strata of the Southern Uplands. There is a marked scenic contrast between the pleasant, intensely cultivated and well-wooded lowlands on the drift-covered, softer Devono-Carboniferous and Carboniferous sedimentary rocks to the north-west on the downthrow side of the fault, and the bleaker grass and heather moorland of the Lammermuir Hills. This higher ground is formed of older, harder Ordovician and Silurian strata, folded and indurated by the Caledonian Orogeny. Carboniferous igneous rocks, being more resistant to erosion than the surrounding sedimentary rocks with which they are associated, form conspicuous features of the lowland landscape. The Garleton Hills, Kilduff Hill and Whitelaw Hill consist of trachytic lavas, Kingston Hill of basaltic lavas, North Berwick Law is a trachytic plug and Traprain Law a phonolitic laccolith.

The low-lying coastal lands of the district attracted early agricultural communities and there were settlements at Chester and Traprain before Roman times. Agriculture has been the main industry throughout the centuries, its prosperity suffering intermittently from economic depressions and, in mediaeval times, from the ravages of invading English armies. Farming techniques began to improve radically in the 18th century and today there is a prosperous agricultural industry with grain and potatoes grown on the very fertile lower ground and sheep-grazing on the higher ground of the Lammermuir Hills. In the western part of the district, coal mining was of importance as far back as the 13th century but this activity has all but ceased.

The main centres of population are the old county town of Haddington, the mining towns of Prestonpans and Tranent and the small ports of Cockenzie, Port Seton and North Berwick. Rural villages and farmsteads are scattered throughout the district. The geography, history and economic activities of the district are discussed in the East Lothian volume of the *Third Statistical Account of Scotland* (Snodgrass, 1953).

GEOLOGICAL HISTORY

The oldest rocks of the district are Ordovician greywackes and shales which contain graptolite faunas indicating ages from early Caradoc (or even late Llandeilo) to Ashgill. The deposits were laid down in the Iapetus Ocean, an extensive trough or geosyncline aligned in a north-east direction. Periodic turbidity currents swept into the trough depositing the greywackes and associated siltstones and shales in fairly deep water along the trough margins, while cherts and graptolitic shales slowly accumulated in the abyssal environment of the deep ocean. Contemporaneous volcanic activity is represented by a felsic acid lava and a silicified tuff associated with the turbidites and tuffaceous bands in the abyssal deposits. Geosynclinal deposition of turbidites, and abyssal sediments continued throughout most of the Silurian, though graptolite evidence indicates that only rocks of Llandovery age are preserved in the district. Earth movements of the Caledonian Orogeny, bringing about closure of the Iapetus Ocean, were active at intervals throughout the geosynclinal deposition, and reached their climax towards the end of the Silurian. Subduction along the Iapetus suture intensely folded and thrust-faulted the Ordovician and Silurian strata along the generally NE–SW Caledonian trend. Within accretionary prisms, turbidite and deep-water deposits were brought into close structural proximity.

A long period of erosion followed the mountain-building before deposition resumed in the Lower Devonian. Torrential conditions in the mountainous terrain produced coarse Lower Devonian conglomerates resting with pronounced unconformity on the eroded Ordovician land surface. Numerous, mainly acid, minor intrusions and small granodiorite masses were intruded into the Ordovician and Silurian strata, probably during Lower Devonian times. The final phase of the Caledonian Orogeny brought to an end Lower Devonian deposition and initiated or reactivated the major Southern Upland Fault-belt including the Lammermuir Fault and the Dunbar–Gifford Fault. Sedimentation resumed late in the Upper Devonian with fluviatile and lacustrine deposition of the red, mainly sandstone, beds of

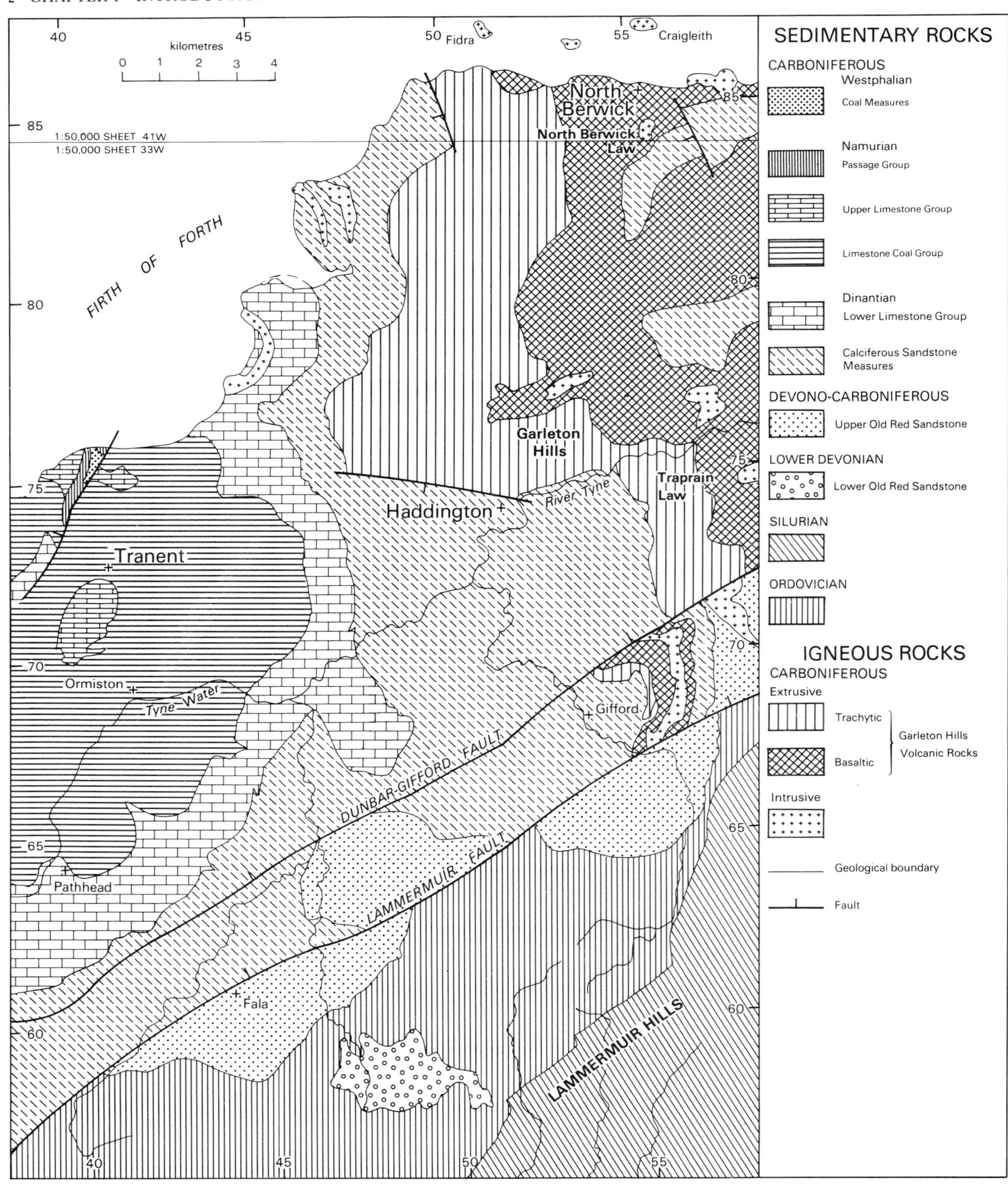

Figure 2 Generalised map of the solid geology of the district

the Upper Old Red Sandstone and probably continued into Carboniferous times. South of the Lammermuir Fault these beds lie unconformably on eroded Ordovician rocks, with an impersistent basal conglomerate. Cornstone beds and nodules are indicative of soils formed in semi-arid conditions.

The appearance of argillaceous beds of Carboniferous type is taken as the base of the succeeding Calciferous Sandstone Measures. Deposited in a shallow, lagoonal, non-marine environment the lower strata are mainly composed of greyish and reddish mudstones, siltstones and sandstones with cementstone ribs. Thick lenticular sandstones were deposited in meandering river-channels. Brief periods of volcanism, indicated by several thin tuff bands, were precursors of the major period of volcanic activity which produced the Garleton Hills Volcanic Rocks, approximately contemporaneous with the eruption of the Arthur's Seat volcano to the west. Volcanic activity was first centred in the north of the district where the basaltic tuffs are thickest, but later moved to the Garleton Hills where the basaltic and trachytic lavas reach their maximum thickness. The lavas were erupted under mainly subaerial conditions, but bedded tuff and agglomerate beds show evidence of deposition in shallow water and re-sorting by wave-action. Following the volcanic episode, the cyclic sedimentation typical of the Scottish Carboniferous was established. Marine incursions commenced and became more frequent towards the close of the Calciferous Sandstone times, while periodic development of swampy tropical forests gave rise to the earliest thin coal seams. The acme of marine conditions occurred during the deposition of the Lower Limestone Group. There the well-marked rhythm of recurring cycles is interpreted as follows—subsidence of the area below a shallow sea, allowing formation of thick marine limestone; subsequent shoaling of the sea by accumulations of mud, silt and sand; build-up of a fresh land surface by deltas of rivers probably flowing from the north-west; growth of tropical forests on the swampy land, with formation of deep soils; finally inundation by the sea to start another cycle.

Shallower water, deltaic environments prevailed during the deposition of the Limestone Coal Group. Conditions were particularly favourable for the growth, accumulation and preservation of vegetation to form thick coal seams. No limestones are present and the evidence of marine conditions is restricted to two marine bands and several bands with *Lingula* only. Limestone deposition was resumed in Upper Limestone Group times but the environment was predominantly deltaic, producing thick sandstones and thin coals. The rocks locally were uplifted and some strata eroded prior to the deposition of the Passage Group. Dominantly fluviatile conditions existed during these times, giving thick, coarse-grained pebbly sandstones. Unstable periods of alternating deposition and erosion produced unconformities with conglomerates resting on eroded surfaces. Marine incursions are indicated by two marine bands in the lower part, while later more terrestrial conditions, with emergence of the land, are shown by thin coals. Red-staining of some strata occurred, but its age and origin are obscure.

At the start of the Lower Coal Measures, conditions returned to those prevailing during Limestone Coal Group times. Thick coal seams indicate prolonged accumulation

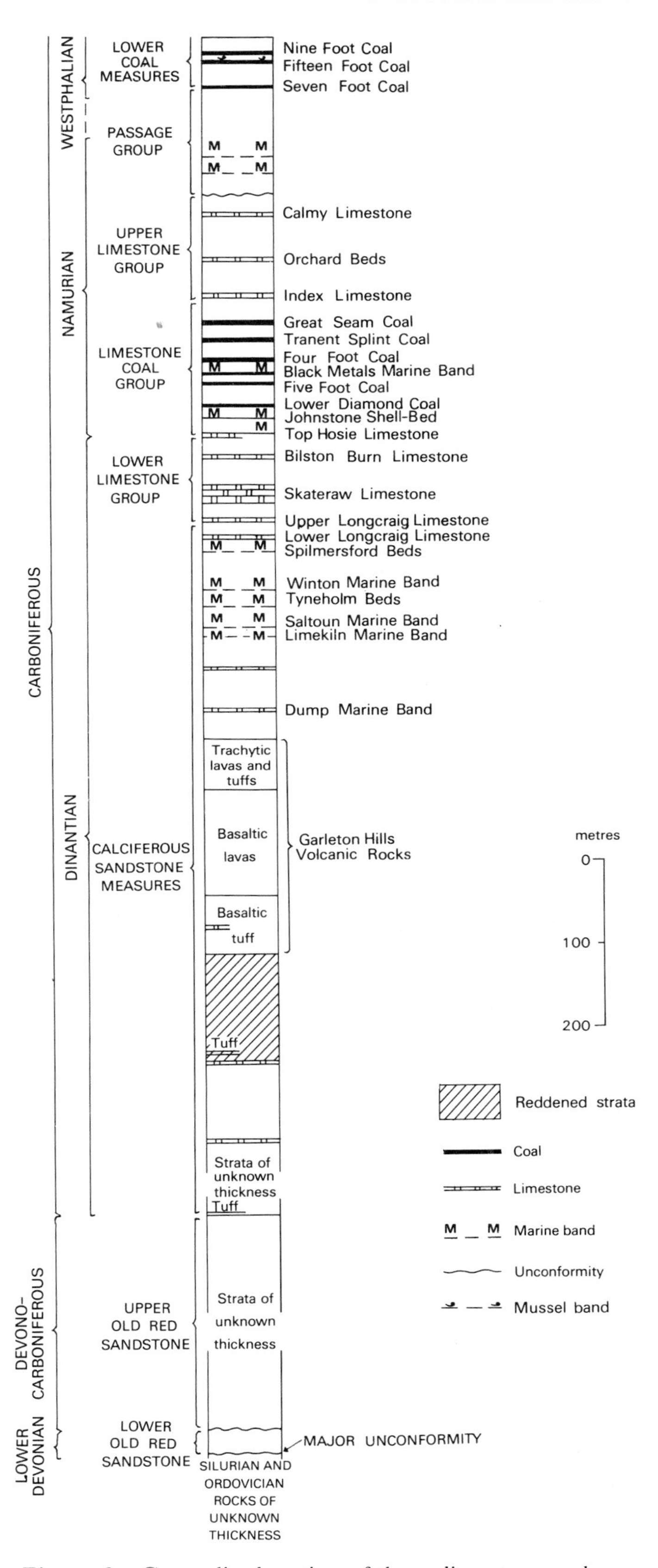

Figure 3 Generalised section of the sedimentary and volcanic rocks

under swamp conditions. Almost certainly, a thick development of Coal Measures strata, similar to that of the Midlothian Coalfield to the west, was deposited in the district. Subsequent erosion, however, has removed all but a basal part of the Lower Coal Measures.

Igneous intrusions occurred periodically through the Carboniferous. Associated with the Garleton Hills Volcanic Rocks are agglomerate- and trachyte-filled necks, particularly abundant round North Berwick, and late-stage sills, dykes and laccoliths of olivine-dolerite, teschenite, basanite and trachyte. Small cryptovents affect strata at least as high as the Passage Group, but may be part of late-Carboniferous episodes of intrusions which produced W–E-trending quartz-dolerite dykes and olivine-dolerite and teschenite sills.

Earth movements, intermittent throughout the Carboniferous, reached their climax in the Hercynian Orogeny of the late-Carboniferous or Permian. The Carboniferous and Devono-Carboniferous rocks were folded into broad shallow synclines and anticlines containing numerous gentle minor folds and many small faults. In the west, the East Lothian Coalfield lies in a basin and to the north-west the Garleton Hills Volcanic Rocks form gentle folds with a north-east axis. The Dunbar–Gifford Fault and the Lammermuir Fault were again active.

For the great interval of time between the Hercynian Orogeny and the Pleistocene, the little evidence of the geological history comes from outwith the district. Much of the time the district was land, subject to erosion, which by Tertiary time had evolved an outcrop pattern not unlike the present. Any sediments deposited in this have been subsequently eroded.

During the Pleistocene, several glaciations occurred which were separated by milder interglacials. During each glaciation the district was completely covered by a thick ice-sheet, which flowed in an ENE direction, eroding the rocks, particularly the softer Carboniferous sediments, and moulding the underlying topography in the direction of ice-flow. Each ice-sheet laid down an extensive ground moraine of till (boulder clay) and infilled the pre-Glacial valleys, but most deposits are likely to belong to the latest (Devensian) glaciation. During deglaciation, the Lammermuir Hills experienced periglacial conditions, in which thick solifluction deposits formed. Meltwater from the decaying ice cut prominent glacial drainage channels on the northern slopes of the Lammermuir and Garleton hills, and deposited mounds, ridges and terraces of sand and gravel.

Fluctuations in the volume of the ice-sheets produced changes in world sea-level and caused differential isostatic movement of the land. Interaction of these effects produced changes in the local sea-level. In late-Glacial times, sea-level in the district reached a maximum of 35 m above OD, forming terraces which tilt eastwards, and marine sediments were deposited. As sea-level fell, lower terraces were cut at approximately 24 and 15 m above OD. Subsequently the sea fell well below its present level. With the post-Glacial (or Flandrian) transgression sea-level rose again to 8 m above OD. Remaining static for about two thousand years, the sea formed wave-cut platforms with littoral sand and shingle deposits, which are backed by cliffs along the steeper coasts, and wide estuarine flats at Aberlady Bay. As the sea has fallen to its present level extensive areas of blown sand dunes have formed on and behind the main raised beaches.

Inland, the milder, wetter conditions following glaciation encouraged growth of hill-peat on the Lammermuir Hills and basin-peat at lower levels. Rivers have formed terrace- and floodplain-alluvium, and the numerous lakes, left in hollows afer glaciation, have mostly been infilled by laminated silt, clay and peat. Made-ground includes reclamation of the intertidal zone at Cockenzie, large coal bings, and major road and rail embankments. ADM, WT

CHAPTER 2

Ordovician and Silurian

INTRODUCTION

Ordovician and Silurian strata crop out in the southern and south-eastern parts of the district, where they occupy an elevated tract of country forming the south-west end of the Lammermuir Hills. The high ground underlain by these relatively hard rocks is bounded to the north-west by the Lammermuir Fault which brings down, on its north-west side, generally less resistant strata of Carboniferous and Devono-Carboniferous age.

The Ordovician rocks are partially obscured by the Carfrae and Kelphope outliers of Lower Devonian (Lower Old Red Sandstone) conglomerate and by Devono-Carboniferous (Upper Old Red Sandstone) sediments of the Fala and Longyester outliers lying adjacent to the Lammermuir Fault (Figure 4).

Lower Palaeozoic strata are fairly well exposed in the many streams, particularly in the eastern and central areas. Relatively few exposures are seen on the smooth, rounded, grass-covered hillsides where the bedrock is overlain by a layer of broken rock debris, probably formed by periglacial action in late-Glacial times.

The north-west part of the Lower Palaeozoic area is occupied by rocks of Ordovician age while Silurian strata crop out in the south-east corner. The position of the boundary between the two systems on the current edition of the 1:50 000 Haddington Sheet differs from previous published maps in that it is placed about 1 km farther north. The revised boundary, which is drawn on faunal evidence alone, is poorly defined and may well be a faulted junction.

The strata consist for the most part of alternations of greywackes, grey siltstones, shales and mudstones, together with minor occurrences of chert and black shales in the Ordovician rocks. Some thick developments of flaggy and

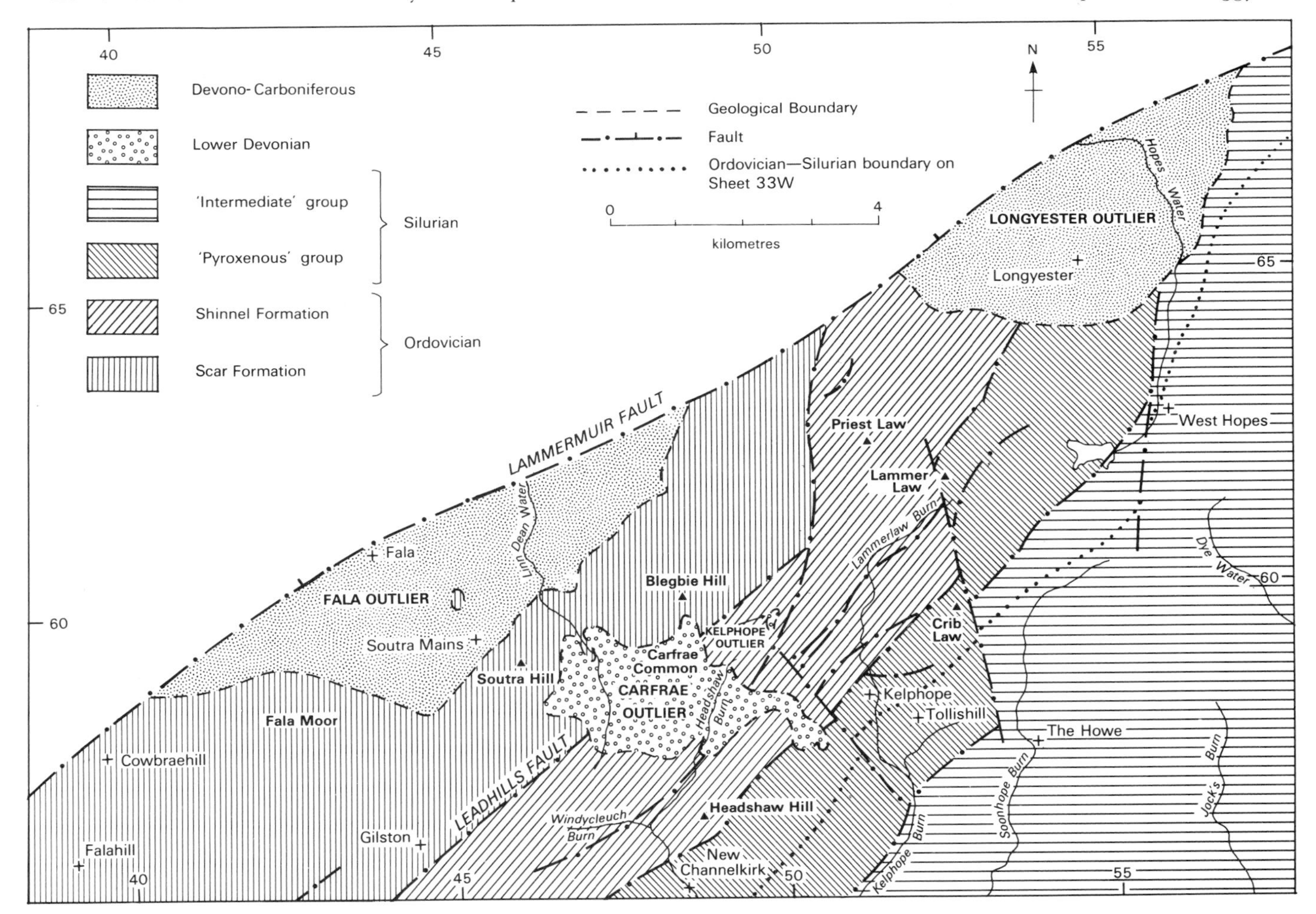

Figure 4 Generalised map of the Lower Palaeozoic rocks

thinly-bedded greywackes and siltstones can be traced for a few kilometres along strike but generally the succession is fairly uniform and it is not possible to correlate beds or groups of beds between isolated exposures. For the most part, the strata strike NE to SW, although in the vicinity of Lammer Law, the strike swings to a more north–south direction. Dips are commonly steep and often vertical. At some localities, sedimentary structures reveal a considerable alternation in younging direction between north and south.

Graptolites have been collected from numerous localities in the Lower Palaeozoic rocks. In the Ordovician, the graptolites are found mostly in the black shales which stratigraphically underlie the greywacke succession although a few specimens have been collected from dark shales interbedded with the greywackes. Undoubted Silurian faunas were obtained from only two localities, in both cases from thin, dark shales interbedded with the greywackes.

The broad structural pattern involves strike-trending, fault-bounded blocks which, overall, contain successively younger sequences towards the south-east. Within each block, however, the strata generally young towards the north-west.

LITHOLOGY

The Lower Palaeozoic succession consists of a thick sequence of turbidite deposits composed predominantly of greywackes, siltstones, shales and mudstones with minor amounts of chert and black graptolite-bearing shale. A few thin beds of tuff or tuffaceous mudstone occur interbedded with black shale and chert in the Lammerlaw Burn, while a decomposed red-brown, silicified rock, possibly a rhyolitic ash-flow, crops out in a tributary of the Kelphope Burn.

The greywacke beds range in thickness from a few centimetres up to about 9 m but are usually less than 2 m. They often display some grading with a thin, coarse, sand base, a thin, silty top and the main thickness of the bed composed of structureless, medium-grained sand. Bases are normally sharp and sometimes erosive into the underlying unit. Successive beds of greywacke are often separated by a thin, dark mudstone which probably represents the background, pelagic sedimentation.

Grey, thinly laminated siltstones and shales, in units up to 5 m thick, are quite common. Pebbly bands are also of frequent occurrence, especially in the thicker greywacke beds, and usually consist of rounded pebbles of quartz, black shale and chert. A conglomerate exposed in Holmes Quarry near New Channelkirk contains rounded pebbles of igneous rocks up to 0.1 m in diameter as well as angular fragments of black shale up to 0.3 m long. A graptolite fauna referred to the *Dicranograptus clingani* or possibly *Climacograptus wilsoni* Zone of the Lower Hartfell Shales has been collected from the latter.

The relative proportions of greywacke to siltstone and mudstone vary from place to place with greywacke forming between 50 and 90 per cent of the thickness.

Sedimentary structures such as flutes, grooves, transverse and longitudinal ripples, cross-bedding and grading have all been observed and used with varying degrees of reliability to determine the order of succession of the beds.

The minor amounts of chert and black shale which occur at the base of the Ordovician succession represent the ocean floor sediments laid down beyond the margins of the submarine fans built up by turbidity currents.

STRATIGRAPHY

Peach and Horne (1899; *in* Clough and others, 1910) placed the Lower Palaeozoic strata of the Lammermuir Hills into the geographical divisions which they called the Northern Belt and the Central Belt. The Northern Belt, consisting of the Ordovician rocks, was thought to include strata of the Caradoc, Llandeilo and, in part, the Arenig series. Subdivision of the Ordovician on the 1910 edition of Sheet 33 was restricted to the representation of small elongate inliers of black shale and chert. The black shales were correlated with the Glenkiln and Hartfell shales of the Moffat district by means of their graptolite faunas.

The inclusion of the Arenig Series was based on the presence of cherts, which in some sections at Ballantrae and Leadhills are associated with volcanic rocks and black shales containing Arenig graptolites. Subsequent work by several authors (Lamont and Lindstrom, 1957; Kelling, 1961; Walton, 1961) has established that in other localities the association of cherts with volcanic rocks and black shales is of Llandeilian or Caradoc age. The proven occurrences of Arenig strata are restricted to areas north of, and including, the Leadhills black shale belt. Since the present district is well to the south of the strike extension of this belt, and the oldest graptolite fauna in the associated black shales is referred to the *Nemagraptus gracilis* Zone, there is no evidence to suggest that the cherts in this area are older than Llandeilian.

Ordovician rocks exposed in the upper reaches of the Lammerlaw Burn contain graptolite faunas ranging from the *N. gracilis* Zone to the *Dicellograptus anceps* Zone. The Glenkiln and Lower Hartfell faunas occur in black shales associated with beds of chert and tuffaceous mudstone. In contrast, the Upper Hartfell faunas occur in dark shales interbedded with a thick greywacke and siltstone succession considered by Peach and Horne (1899) to be equivalent to the Lowther Shales.

This illustrates the well-documented change in the character of parts of the Lower Palaeozoic succession when traced across strike from the Central Belt to the Northern Belt, since the Upper Hartfell Shales at Dobb's Linn, Moffat, some 56 km to the SW, consist of a thin succession of grey and black mudstones. A similar situation prevails in the lower part of the Silurian succession where the thick sequence of greywackes in the northern part of the Central Belt in this area, named the Queensberry Grits by Peach and Horne (1899), are represented at Dobb's Linn by the 43 m of black and grey mudstones of the Birkhill Shales. A sparse fauna of fragmentary monograptids collected from exposures in the Dye Water include types common in the *Pristiograptus gregarius* Zone of the Middle Birkhill Shales. At another locality in Jock's Burn near the south-east corner of the area, a graptolite fauna including *Rastrites sp.* was collected from dark partings in a greywacke succession, proving the presence of strata equivalent to the highest Birkhill Shales or lowest Gala Group of Dobb's Linn.

During the resurvey it was found that lithological subdivision of the Lower Palaeozoic rocks in the field was impractical and the boundary between the Ordovician and the Silurian was determined on graptolite evidence alone. Peach and Horne placed the boundary between the two systems at a lithological change, about 1 km to the south of the newly proposed palaeontological boundary. The line formerly chosen was the boundary between the Lowther Shales and the Queensberry Grits, but the graptolite evidence collected during the resurvey indicates that Silurian rocks occur to the north of this line. This may be due to structural repetition, but there are insufficient structural or palaeontological data to confirm or refute this possibility.

Subsequent petrographic examination of the Northern Belt greywackes in this area has revealed the presence of lithostratigraphic units which can be correlated with the Scar and Shinnel formations of Nithsdale (Floyd, 1982) some 85 km along strike to the south-west. A similar petrographic correlation may be made between greywackes in the Central Belt of this area and the Pyroxenous and Intermediate groups of the Peebles area (Walton, 1955). These formations are shown in Figure 4, where, by analogy with these other areas, it could be considered that the Ordovician–Silurian boundary should be placed another 1 km further north to lie along the Pyroxenous group–Shinnel Formation boundary.

STRUCTURE

According to Lapworth (1889), the structure of the Lower Palaeozoic rocks of the Southern Uplands consists of numerous isoclinal folds which have been refolded to form a broad open anticlinorium and synclinorium pair. The anticlinorium, called the Leadhills Endocline, lies to the north-west of the synclinorium which was called the Hawick Exocline. The axes of these broad open folds were named the Leadhills and Hawick lines respectively. Peach and Horne (1899) accepted Lapworth's interpretation of the structure and extended it throughout the outcrop of the Lower Palaeozoic rocks. In the present district, the Ordovician and Silurian rocks lie just south of the Leadhills Line and towards the northern edge of what was supposed to be the common limb of these two broad folds.

Lapworth's views on the structure of the Lower Palaeozoic rocks were abandoned from about 1955 onwards largely as a result of advances in sedimentological research. It was found that sedimentary structures, which are a feature of turbidite deposition, could be used to determine the direction of younging at individual exposures rather than having to rely on an inferred order of succession deduced from the infrequent fossils.

A general structural pattern has emerged of fault-bounded blocks containing steeply-dipping successions which young towards the north-west within any one block. This, however, is more than offset by the bounding strike-faults which have a considerable relative upthrow to the north-west, with the result that, when traced across strike from north-west to south-east, each successive block contains progressively younger strata. Several workers (Craig and Walton, 1959; Kelling, 1961) have described a monoclinal structure within some of the fault-bounded blocks, with belts of repeatedly folded strata disposed alternately with belts of largely unfolded but steeply-dipping strata. Structural elements which may be attributable to this monoclinal structure have been noted in the district but any monoclines would be essentially of secondary importance, the overall structure being dominated by the fault-bounded blocks of steeply-dipping strata.

Recently it has become clear that this structural pattern may be explained using an accretionary prism model for the Southern Uplands (McKerrow and others, 1977). In this model, it is envisaged that during subduction of a southern Iapetus oceanic plate northwards under the Laurentian continent, faulted packets of sediments, originally deposited in the intervening trench, were discontinuously accreted on to the continental margin. Continued convergence of the crustal plates would lead to the build up of a prism of progressively-rotated, imbricated, fault-bounded blocks containing steeply-dipping, northward-younging strata. The northernmost and earliest accreted block would contain the oldest succession and be adjoined southwards by later accreted blocks containing successively younger sequences.

In the course of the present resurvey, the order of succession of the rocks was determined, as far as possible, by means of sedimentary structures. At many exposures, however, particularly in the Ordovician, the strata were found to be greatly sheared and indurated, and any bottom-surface structures were difficult to detect. At some localities graded bedding was used as an indication of the direction of younging.

The strike of the strata is mainly north-east to south-west and dips are steep or vertical. Notable variations in strike occur in the area north-west, west and south-west of Lammer Law where a more north–south direction is prominent.

It has been found difficult to make a detailed interpretation of the structure within the Lower Palaeozoic rocks. Exposures are confined largely to stream sections, which give only a discontinuous picture of the tectonics. In addition, outcrops of shales with identifiable graptolite faunas are concentrated at a few localities, and, in the Ordovician, it is not known which graptolite zones are represented by the greywacke-shale turbidite succession which occupies most of the area.

The overall structure of the Ordovician in the area is thought to be dominated by several strike-trending thrust faults closely associated with lenticular outcrops of chert and cherty shale which are probably among the oldest rocks exposed in the area. The conjectural position of these faults is indicated on Figure 4.

DETAILS

Strata immediately to the north-west of one of these dislocations are seen in Lammerlaw Burn [5127 6050] where there are exposures of sheared and fractured cherts, shales, mudstones and greywackes, probably representing a zone of imbrication. A prominent feature on the south-east slopes of Headshaw Hill [489 563], extending to the north-east for about 2 km, is thought to represent the outcrop of a large, reverse fault. Debris of black, cherty shale and green, grey and black chert is seen at several places near the base of the feature.

Plate 1 Thrust-faulting in Ordovician shales and greywackes, Holmes Quarry, Channelkirk (D3021)

The rocks to the north-west include thick beds of greywacke, whereas to the south-east the strata are predominantly flaggy and thinly-bedded siltstones, shales and mudstones, forming a zone of low relief.

A thrust dipping to the NW at 40° is exposed on the east bank of the Linn Dean Water [4653 5982], about 1.25 km ENE of Soutra Mains. The strata above and below the dislocation are alternating beds of greywacke, siltstone and mudstone and the quartz-veined fault breccia is about 0.4 m thick. About 4.7 km to the south, a small thrust dipping north at 36° is exposed in Holmes Quarry [4842 5548] near New Channelkirk, where shales and siltstones are thrust southwards over greywacke (Plate 1). On the east bank of the Kelphope Burn [5125 5870], 340 m upstream from Kelphope, a thrust is seen dipping north at 35° in sheared and shattered vertical shales and siltstones.

The conjectural positions of three faults trending almost normal to the general strike are shown on Figure 4. These are possibly wrench faults with a sinistral movement. The fault about 700 m SW of Kelphope [5080 5785] and the dislocation east of Crib Law [5280 5957] appear to displace the outcrops of chert and black shale, while the fault near West Hopes [5560 6255] displaces the Ordovician – Silurian boundary.

Gently plunging minor folds are seen at several localities where the predominant style is asymmetrical rather than isoclinal. Several minor folds were noted in old quarries on the north slopes of Soutra Hill [4600 5935] where the beds generally strike north-east to south-west, are vertical or dip steeply to the south-east, and young towards the north-west. In the central part of the quarries, a tight, minor syncline occurs with a vertical north limb, a gently dipping south limb and a gentle plunge to the north-east. A tight anticline, overfolded to the south-east and plunging gently to the north-east, is exposed at the head of Colzie Cleugh, 240 m NE of Tollishill [5204 5810). Steeply plunging minor folds also occur, as in the Windycleuch Burn [4747 5648] where two asymmetrical minor folds can be seen plunging ENE at 45° – 50° with the north-west limbs the more steeply inclined.

In the Silurian rocks, the detailed structure is not known due to poor exposure and lack of marker horizons. The overall structure is probably similar to that in the Ordovician with large strike faults separating blocks of steeply dipping strata. The graptolite faunas suggest that younger beds come on to the south-east but numerous reversals of younging direction indicate that the structure is complicated in detail. Vertical cleavage trending north-east to south-west in mudstone was recorded in the Longformacus Burn [5409 5869] and in the area north of The Howe [5381 5791].

PALAEONTOLOGY

The following section is based on the identifications of the graptolites by Dr I. Strachan.

Almost all the graptolites from the district belong to the Ordovician. There is one definite Silurian locality in the south-east corner [5668 5586] where a small fauna includes *Rastrites sp.* of a type found in the highest Birkhill Shales and lower Gala Group. A larger collection from the headwaters of the Dye Water [5664 6040] includes fragmentary monograptids of types common in the *Pristiograptus gregarius* Zone of the Middle Birkhill Shales. This latter locality is very close to the boundary of the Ordovician and Silurian given on the 1910 edition of Sheet 33 but the occurrence of shales with a Birkhill fauna some 3 km to the SW (in the Soonhope Burn) recorded on that map has not been satisfactorily confirmed by recollecting. The old collections include a very doubtful monograptid stipe fragment and the new material has only a *Climacograptus sp.* recognisable whose age is not diagnostic. Farther north, on the north slopes of Lammer Law, one locality [5261 6224] has yielded *Diplograptus sp.* of *D. modestus* Lapworth type, a species more characteristic of the lowest Birkhill than the Upper Ordovician, but about 500 m NE along the strike another locality is clearly in the Upper Ordovician [5303 6255] while to the south-east there are numerous localities in the Kelphope Burn yielding Ordovician faunas.

A notable feature of the Ordovician of the district is the comparative abundance of localities with Upper Hartfell faunas compared with their absence farther east. Unfortunately the Longyester Outlier of Devono-Carboniferous rocks lying to the south-east of Gifford obscures any chance of tracing the beds directly along strike.

The large number of Ordovician localities (about 15) in the upper reaches of the Kelphope Burn (mainly in National Grid square 51 60) yield faunas ranging in age from Glenkiln (*N. gracilis* Zone) to Upper Hartfell (*D. anceps* Zone) but do not show any clear pattern of ages. A further series of 9 localities about 2 km to the SW [50 59] similarly shows no regular plan and it is probable that these all represent repetition by folding and/or faulting of a comparatively thin sequence.

A comparable situation is found again in the Headshaw Burn [47 56] where, although most of the localities contain Upper Hartfell faunas, there are some lower beds present, not obviously in the cores of anticlines. A kilometre farther south, at New Channelkirk, two localities yielded Lower Hartfell or Upper Glenkiln faunas with no trace of higher beds.

The faunas are mostly in black shales and are poorly preserved as a result of tectonism but it is possible to recognise all the Hartfell Shales zones in one place or another and it seems that the Ordovician succession is complete. The Silurian faunas are too poor to be able to make a similar statement about the Llandovery but in any case, the Birkhill Shales of Moffat are here represented by a much thicker sequence of coarser beds. JDF, WT

CHAPTER 3

Lower Devonian (Lower Old Red Sandstone)

In the district, strata of Lower Old Red Sandstone facies, assigned to the Lower Devonian, occur only in the Lammermuir Hills to the south-east of the Southern Upland Fault. Two outliers are present, in which the beds rest with great unconformity on an eroded surface of vertical or highly inclined Ordovician strata.

The larger, Carfrae Common Outlier, 4 km from west to east and 2 km from north to south, lies south-east of Fala in the Huntershall–Carfrae Common area between Soutra Hill, Kelphope Hill and Headshaw Hill (Figure 5). The strata in this outlier consist of fine-, medium- and coarse-grained conglomerates with lenticular bands of red-brown, pebbly sandstone, and thin beds of red-brown siltstone and mudstone. The conglomerates were laid down as torrential gravels in a fluviatile or lacustrine environment. In the west, along the Linn Dean Water, the beds are more or less flat-lying or dip very gently southwards, while gentle dips to the north-west occur in the middle of the outlier in the upper part of the Headshaw Burn. Red-brown conglomerate is exposed in the eastern part of the outlier in the headwaters of the Hillhouse Burn.

The Kelphope Outlier, an additional small area of conglomerate, was found in the course of the present revision north of Kelphope Hill at the head of a small tributary of the Kelphope Burn.

The boulders in the conglomerates are well rounded, with a maximum diameter of about 0.5 m. They are mainly derived from the Lower Palaeozoic rocks of the Southern Uplands and are predominantly of greywacke, with scattered boulders of chert, quartzite, and igneous rocks including

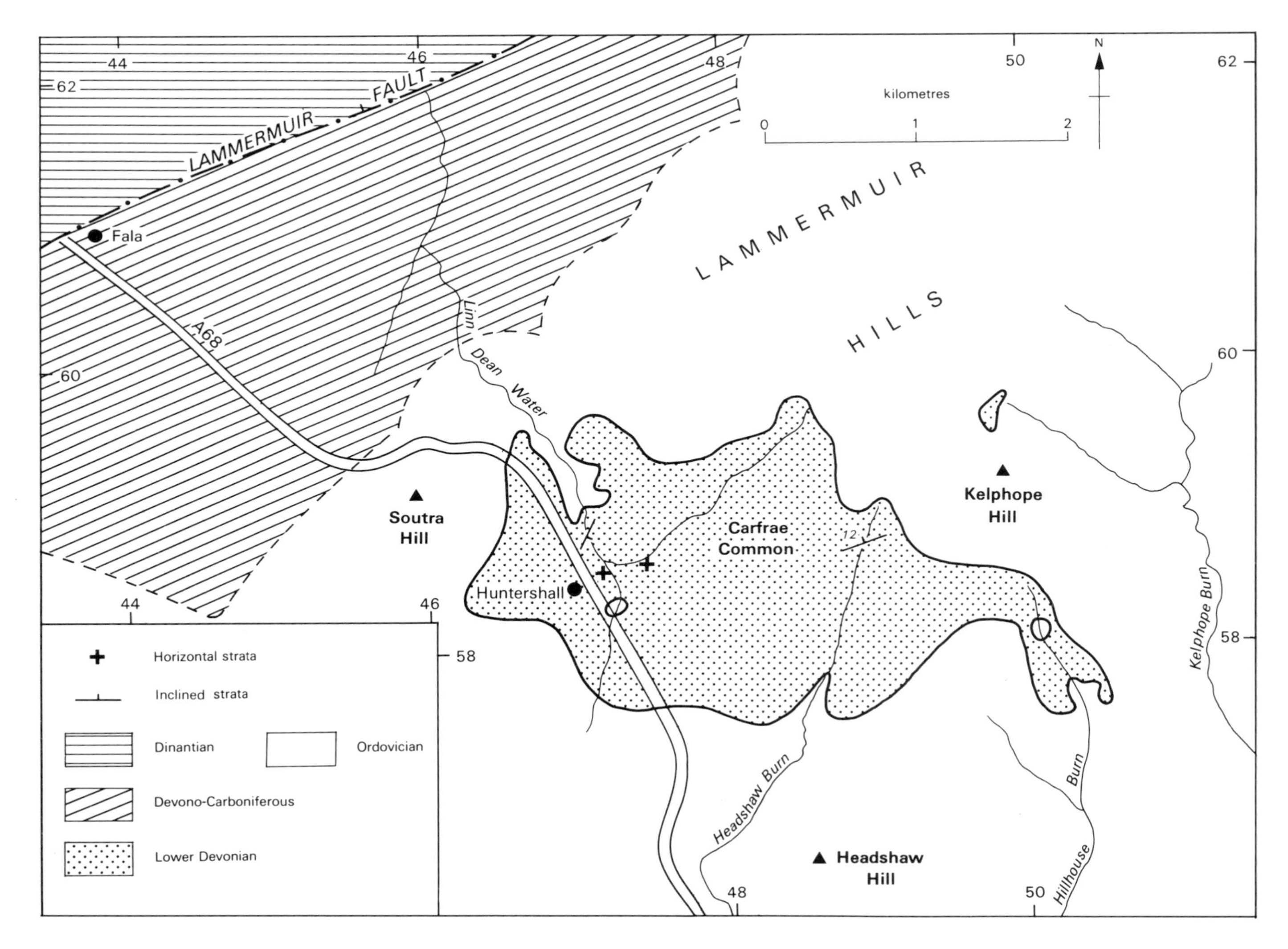

Figure 5 Generalised map of the Lower Devonian outliers

quartz-porphyry and felsite (Plate 2).

No fossils have been found in these beds to indicate their age. On previous editions of Sheet 33 the strata were shown as Upper Old Red Sandstone (Geikie *in* Howell and others, 1866; Bailey *in* Clough and others, 1910). The Carfrae Common Outlier was thought to be an eroded remnant of the Great Conglomerate (Bailey *in* Clough and others, 1910, p. 30) which occur some 20 km to the east in the Dunbar District (Sheet 33E), extending southwards from Spott across the Lammermuir Hills. The Great Conglomerate is now considered to be of Lower Devonian age (Davies and others, in preparation). As the conglomerates in the Carfrae Common Outlier are lithologically similar, they have also been assigned to the Lower Devonian.

Lower Devonian deposition came to an end with the onset of important earth movements which marked the final phase of the Caledonian Orogeny. WT, ADM

Plate 2 Lower Devonian conglomerate, predominantly greywacke cobbles, Linn Dean Water (D3023)

CHAPTER 4

Devono-Carboniferous (Upper Old Red Sandstone)

INTRODUCTION

Strata of Upper Old Red Sandstone facies in south-east Scotland, from fossil evidence, are believed to include beds of both Upper Devonian and Lower Carboniferous age. Hence they are classified as Devono-Carboniferous.

Five areas of Upper Old Red Sandstone strata occur in the district (Figure 6). Immediately south of the Lammermuir Fault there are two outliers, the Fala Outlier and the Longyester Outlier. Both outliers rest with marked unconformity on an eroded land surface of steeply-dipping Ordovician strata. There is a basal conglomerate in the latter one. The other three areas lie between the Dunbar–Gifford Fault and the Lammermuir Fault, the small Costerton Inlier, the larger Humbie Inlier and separated from it by the Gifford Basin, an area in the east around Carfrae. In all three of these areas the Upper Old Red Sandstone is overlain conformably by Dinantian strata. Devono-Carboniferous strata may also have been proved at the base of the IGS Spilmersford Borehole, below a faulted contact with Dinantian strata (Davies, 1974, p. 7).

The areas of Upper Old Red Sandstone strata are generally obscured by a thick mantle of till and glacial sand and gravel. Exposures are restricted to excellent sections in the deeply incised valleys, many cut or deepened by glacial meltwater, which cross the outcrop areas. In the west the main exposures are in the valleys of the Keith Water, Humbie Water, Birns Water and their tributaries. It is valleys in the headwaters of the Gifford Water and Whittingehame Water which provide sections in the Longyester Outlier and at Carfrae.

ENVIRONMENT

The Upper Old Red Sandstone sediments are partly fluviatile and partly lacustrine in origin. Conglomerates accumulated in piedmont alluvial fans which extended northwards from high ground to the south. Sandstones, siltstones and silty mudstones were laid down on an alluvial plain. Describing the Upper Old Red Sandstone of the Burnmouth, Duns, Greenlaw and Kelso areas in Berwickshire and Roxburgh across the Lammermuir Hills to the south-east, Smith (1967) concluded that the sandstones were deposited in meandering river channels flowing through floodplain areas, while Leeder (1973) thought that the Upper Old Red Sandstone of the southern Scottish Borders is a wholly fluviatile succession.

Cornstones in the succession are regarded as pedogenic carbonates which are the calcareous horizons in fossil soils. Burgess (1960) has shown that the profiles of the cornstones in the Upper Old Red Sandstone of south Ayrshire compare closely with the Pliocene caliche limestones of south-central USA, described as pedocal soils formed under semi-arid conditions.

The Upper Old Red Sandstone sediments on the south-east side of the Lammermuir Fault in the Fala and Longyester outliers fill in the upper parts of pre-Upper Old Red Sandstone valleys (Bailey *in* Clough and others, 1910, p. 36).

CORRELATION

The only fossils found in the Upper Old Red Sandstone rocks of the district are plant remains, too poorly preserved to determine their precise age. In the past the beds have been classified as Upper Old Red Sandstone mainly because of their lithological resemblance to rocks of this age, 20 to 30 km to the east on the Berwickshire coast between Cove Harbour and Siccar Point. These strata have yielded scales of *Holoptychius nobilissimus* and a similar scale was found in a fallen block in a river gorge near Whittingehame, beyond the eastern margin of the district (Bailey *in* Clough and others, 1910, p. 35).

In the district the Upper Old Red Sandstone facies passes conformably up into strata of Carboniferous facies. The only marker band used to delineate the boundary is a thin volcanic horizon at the base of the Carboniferous, exposed in the Humbie Water at Humbie Mill (p. 44). Elsewhere the boundary indicated on the map between Devono-Carboniferous and Dinantian is conjectural.

It is likely that the Upper Old Red Sandstone facies persisted into Carboniferous times and that at least part of the Upper Old Red Sandstone strata is in fact of Carboniferous age. Neves and others (1973, p. 51) reported that the oldest miospore zone recorded from the Calciferous Sandstone Measures of the Midland Valley of Scotland is the *Schopfites claviger–Auroraspora macra* (CM) Zone. As this is not the oldest miospore zone in the Carboniferous this implies that the Upper Old Red Sandstone beds, which appear to conformably underlie the Calciferous Sandstone Measures, are in part at least of Carboniferous age, as has been suggested by Waterston (*in* Craig, 1965).

DETAILS

Fala Outlier

The Fala Outlier stretches for 9 km along the south side of the Lammermuir Fault in the Fala Mains–Mavishall area (Figure 6). The outlier is crossed by several north-flowing streams, tributaries of the Fala Dam Burn, the Dean Burn and the Linn Dean Water. It is in the incised valley of the latter two that most exposures occur.

The unconformity between the Upper Old Red Sandstone and the Lower Palaeozoic is exposed in the small Ordovician inlier in

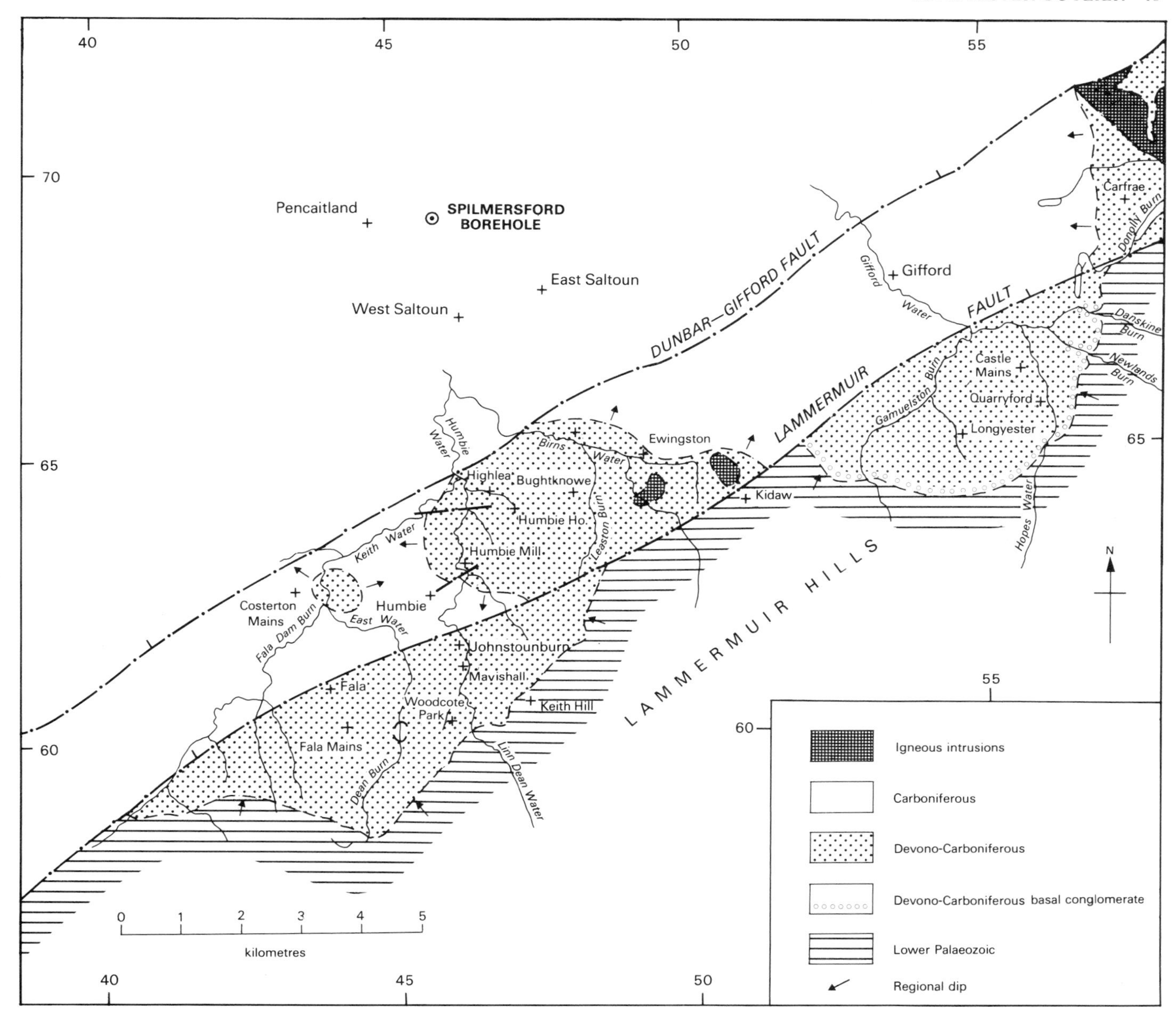

Figure 6 Generalised map of the Devono-Carboniferous rocks

the Dean Burn east of Fala Mains. Two metres of coarse, red-brown conglomerate with well-rounded boulders of greywacke up to 0.4 m across, rest on steeply-dipping Ordovician greywacke and shale. Upstream and downstream from the inlier in the Dean Burn are exposures of the generally horizontal or gently rolling Old Red Sandstone strata (Plate 3). The succession is composed mainly of beds of sandstone, micaceous siltstone and silty mudstone, with a few thin beds of conglomerate. In colour the strata are red and red-brown, with small, green reduction-spots common at some localities. The sandstone beds range in thickness from a few centimetres up to 2 m. Commonly the sandstones are medium- and fine-grained with coarse bands. Thin, pebbly bands with small pebbles of quartz and black, red and grey chert occur widely. Some sandstone beds grade upwards from coarse- to fine-grained with the upper part flaggy and thinly laminated. Cross-bedding, indicating palaeocurrents from the south, are displayed in several exposures. The bases of many sandstone beds are sharp and erosive. The beds of siltstone and silty mudstone vary in thickness up to 5.5 m, but generally include thin beds of sandstone ranging in grain-size from fine to coarse and pebbly.

Strata of similar lithology are exposed farther to the north-east, in the Linn Dean Water and its tributaries around Woodcote Park and south of Mavishall. The general dip in this area is towards the north at low angles. On the south-west bank of the Linn Dean Water, north-east of Woodcote Park, thick siltstone and silty mudstone beds were formerly worked for tile manufacture. Further east, medium- and coarse-grained sandstone with pebbly bands were formerly worked in a small quarry 250 m NW of Keithhill.

Longyester Outlier

The Longyester Outlier lies to the south-east of Gifford. It occupies a semicircular outcrop area, 6 km across, on the south-east side of the Lammermuir Fault in the neighbourhood of Longyester, Quarryford and Castle Mains (Figure 6). The main exposures occur in the east of the outlier in three incised tributaries of the Gifford Water, namely the Hopes Water, the Newlands Burn and the Danskine Burn. Purple, coarse- and fine-grained conglomerates at the base of the succession are exposed in the Danskine Burn, in the Newlands Burn and in the Hopes Water upstream from Quarryford. The basal conglomerates are overlain by massive and flaggy, purple, red, grey and white speckled sandstones with pebbly and conglomeratic bands and subordinate beds of siltstone and silty mudstone. In places the sandstones are cross-bedded. The general dip in this eastern part of the outcrop, is towards the west at angles ranging from 5° to 15°.

Costerton Inlier

This small inlier, less than 1 km across, lies just east of Costerton Mains (Figure 6). Exposures in the Costerton Water show beds of red, purple and brown sandstone, siltstone and silty mudstone. The sandstones are mainly medium- and fine-grained and beds reach a maximum thickness of 3.6 m. Nodules and irregular beds of red and pale brown cornstone occur in the siltstone and silty mudstone bands. The general dip is towards the north-west at low angles, but at some exposures the beds are almost horizontal and undulate gently.

Humbie Inlier

A large outcrop of Upper Old Red Sandstones strata between the Dunbar–Gifford Fault and the Lammermuir Fault forms the Humbie Inlier which extends from Humbie Mill for 6 km to the north-east towards Kidlaw (Figure 6).

The strata in the Humbie Inlier are well exposed in the Humbie Water from its confluence with the Keith Water upstream to Humbie Mill, and in the stream which joins the Humbie Water from the east, upstream from Humbie Mill. Exposures are present in the eastern part of the inlier in the Leaston Burn north-east of Bughtknowe, and in the Birns Water upstream from Ewingston.

Changes in lithology suggest that the Humbie Water section is intersected by at least three faults (Figure 6). The conjectural line of the most northerly dislocation, the Highlea Fault, trending

Plate 3 Alternating bedded sandstones and marls, Devono-Carboniferous, Humbie Water, east of New Mains (D998)

north-east, crosses the Humbie Water west of Highlea. About 330 m to the south, the W–E-trending Church Wood Fault intersects the stream south-west of Highlea. The Humbie Mill Fault, with an ENE trend, crosses the Humbie Water about 140 m upstream from Humbie Mill.

Downstream from Humbie Mill the general dip of the beds is towards the NW at inclinations ranging from 5° to 15°, with minor undulations. From Humbie Mill upstream to the Humbie Mill Fault the strata dip to the SE from 8° to 25°, while south-east of this fault the dip is towards the SW at 10° to 15°. In the Leaston Burn north-east of Bughtknowe and in the Birns Water south-west of Ewingston the strata dip generally south or to the SE at angles ranging from 5° to 17°. Farther upstream in the Birns Water the rocks dip to the NW at 7° to 14°.

Two small basic intrusions cut the Upper Old Red Sandstone strata in the east of the inlier at Ewingston and Kidlaw.

In the Humbie Inlier the Old Red Sandstone strata in general consist of varigated red, purple, brown and green sandstones, silstones and silty mudstones, with, in places, nodules and lenticular bands of cornstone. The succession is made up of a series of lithological units consisting of a bed of sandstone, grading upwards from medium- to fine-grained, overlain by siltstone and silty mudstone with intercalations of silty sandstone. In general the sandstone beds are thin, being on average about 0.6 m thick. An exceptional bed of fine-grained sandstone over 2.4 m thick forms a waterfall in a tributary stream on the east bank of the Humbie Water, west of Humbie House [4636 6394]. The bases of the sandstone beds are commonly sharp and erosive. A few sandstones are coarse-grained in the basal part, with small pebbles of quartz and cornstone.

In the Humbie Water section, the strata north of the Highlea Fault and the strata between the Church Wood Fault and the Humbie Mill Fault are similar in character. Both are composed predominantly of red-brown, deep red and purple-grey medium- and fine-grained sandstone in beds up to 2.5 m thick, with thinner intercalations of variegated red-brown, purple-grey and pale green siltstone and silty mudstone. Poorly-preserved plant remains occur in some of the sandstone beds. Between the Highlea Fault and the Church Wood Fault, however, the succession is characterised by the presence of nodules and irregular beds of cornstone which both occur in the sandstones, and in the siltstones and silty mudstones. A prominent bed of cornstone, 1 m thick, crops out in the core of a gentle anticlinal fold in the east bank of the Humbie Water south-west of Highlea [4624 6411], where the section is:

	Thickness m
Sandstone, purple-grey, fine- and medium-grained, in thin irregular beds with laminae of green silty mudstone	0.13
Cornstone, pale brown and green, sandy	0.13
Cornstone, variegated purple-grey, red-brown and green, irregularly-bedded	1.00
Siltstone and silty mudstone, variegated purple and red with irregular cornstone nodules	0.60

Another bed of red, purple and brown cornstone and hard calcareous sandstone 1.4 m thick, with irregular laminae of purple and green silty mudstone, is exposed on the north bank about 80 m upstream. In the tributary of the Humbie Water immediately south of Highlea are exposures of red, purple, green and brown variegated fine- and medium-grained sandstones, siltstones and silty mudstones with nodules and bands of grey, brown and red cornstone.

In the eastern part of the Humbie Inlier, Upper Old Red Sandstones strata are exposed in the Leaston Burn north-east of Bughtknowe and in the Birns Water upstream from Ewingston. The rocks consist mainly of variegated red, purple, brown and pale green sandstones, siltstones and silty mudstones. The sandstone beds reach a thickness of about 4.5 m and are mainly medium- and fine-grained, with a few pebbly bands containing small quartz pebbles and fragments of siltstone. Some sandstones are soft, argillaceous and friable, while others are hard and siliceous. The siltstone and silty mudstone beds reach a thickness of about 4.5 m, and commonly contain thin beds of sandstone.

The highest beds in the Upper Old Red Sandstone of the inlier are exposed in the Humbie Water and in the tributary stream from the east, south of the Humbie Mill Fault. The rocks are variegated red, purple, brown and green medium- and fine-grained sandstones in beds up to 1.5 m thick with generally thinner bands of siltstone and silty mudstone. Some of the sandstones contain small pebbles of quartz and cornstone, and small nodules of white chert. Nodules and thin bands of cornstone occur throughout.

In the Humbie Inlier the Upper Old Red Sandstone passes conformably up into the Carboniferous. The local upper limit of the Upper Old Red Sandstone is taken at the base of a thin bed of tuff and agglomerate which forms a convenient horizon. This band crops out only in the Humbie Water south of Humbie Mill. The basal Carboniferous strata here consist of variegated purple, red and brown sandstones and purple, red, green and brown siltstones and mudstones with deep red-purple concretions, lithologically very similar to the upper beds of the Upper Old Red Sandstone.

Carfrae

Upper Old Red Sandstone rocks lie between the Dunbar–Gifford Fault and the Lammermuir Fault in the vicinity of Carfrae on the eastern margin of the district. Only the upper beds are seen at outcrop, and the strata dip gently eastwards passing conformably up into the Carboniferous. Large intrusions cut the strata (Figure 6). The rocks are very poorly exposed, except in the south, where horizontal conglomerates and purple-stained, cross-bedded sandstones crop out around the south end of the Donolly Burn Reservoir.

Spilmersford Borehole

The basal 10 m of the Spilmersford Borehole, below a fault at 924 m, are reddened sandstones and conglomerates of Old Red Sandstone facies (Davies, 1974). No fossils were found in these beds, but on lithological similarity they are correlated with the Devono-Carboniferous strata seen only at outcrop south of the Dunbar–Gifford Fault. The lowest 2 m of strata proved in the borehole were pale purple, calcareous sandstones wih calcareous concretions reminiscent of cornstone in the Upper Old Red Sandstone. Cementstone pebbles in the overlying conglomerate, on the other hand, suggest a Lower Carboniferous age. WT, ADM

CHAPTER 5

Lower Carboniferous

INTRODUCTION

Rocks of Lower Carboniferous (Dinantian) age are present over a large part of the district. They are divided into the Calciferous Sandstone Measures and the overlying Lower Limestone Group. The former division comprises a thick sequence of sedimentary rocks but also includes the Garleton Hills Volcanic Rocks. The Lower Limestone Group is relatively much thinner and is entirely composed of sediments including thick limestones. The zonation and classification of the sedimentary rocks are discussed in Chapter 7 and the igneous rocks are described in Chapters 8 and 9.

CALCIFEROUS SANDSTONE MEASURES

In the absence of diagnostic fossils, the base of the Calciferous Sandstone Measures is drawn at the incoming of grey mudstone and other sediments of Carboniferous type overlying the red sandstones of the Devono-Carboniferous. The top of the division is taken at the base of the Upper Longcraig Limestone, which marks the base of the Lower Limestone Group, an horizon which has been correlated over much of central Scotland. The successions in the district and in Midlothian to the west are shown on Figure 7.

The sediments below the Garleton Hills Volcanic Rocks are dominantly grey, brown, red or purple mudstones and siltstones with bands of cementstone (argillaceous dolomite) and grey, green and red sandstones. Thin coals with underlying root-beds show that a cover of vegetation was established at one or two horizons. The environment is interpreted as having been coastal and non-marine, probably lagoonal, and the few fossils recorded indicate non-marine or quasi-marine conditions. The succession indicates the changes produced by the approaching Carboniferous sea as distinct from the foregoing terrestrial, semi-arid conditions of the Old Red Sandstone facies.

The lavas, tuffs and agglomerates of the Garleton Hills Volcanic Rocks were deposited mainly under subaerial conditions although some tuffs and agglomerates show evidence of resorting by water-action.

Following the volcanic episode, the sediments in the upper part of the Calciferous Sandstone Measures indicate the establishment of the conditions which produced the typical cyclic deposition of much of the Scottish Carboniferous succession and the fossils show that marine episodes became increasingly prevalent and of longer duration towards the close of the period. In a typical cycle, a seatearth and coal, indicating subaerial conditions, are followed by marine mudstones indicating a transgression by the sea. These are succeeded by deltaic mustones, siltstones and sandstones, with non-marine fossils, indicating increasing shallowing until subaerial conditions return, with seatearth and coal formation marking the start of the next cycle.

Several thin coals, up to 0.35 m thick, and thick sandstones are present but at the top of the division the Lower and Middle Longcraig limestones, containing rich coral-brachiopod faunas, show that fully marine conditions were established at this time.

About two-thirds of the area north of the Dunbar–Gifford Fault in the district are occupied by sedimentary and volcanic rocks of Calciferous Sandstone Measures age (Figure 8). The rocks have a regional dip to the west under the Lower Limestone Group. On the coast the fairly well-exposed outcrop extends from the edge of the district east of North Berwick to the west and south-west as far as Aberlady. Inland the outcrop, up to 13 km wide, extends southwards by the Garleton Hills, Haddington and Bolton to the Dunbar–Gifford Fault between Bara and Tynehead. Two areas of Calciferous Sandstone Measures strata are present between this fault and the Lammermuir Fault, separated by the Devono-Carboniferous Humbie Inlier. The eastern area, the Gifford Basin, extends from Bara by Gifford to Leehouses and the other area from Humbie to the south-west edge of the district at Tynehead. Exposures inland are good only where the harder volcanic and intrusive rocks of the Garleton Hills Volcanic Rocks form prominent features. Elsewhere the sedimentary rocks are generally drift-covered and exposed only in incised river valleys and old quarries. By far the best section in the Calciferous Sandstone Measures was proved in the IGS Spilmersford Borehole (Davies, 1974) which provided an almost complete sequence (Figure 7). Old boreholes at Samuelston and Lennoxlove gave sections of parts of the succession.

Structure

In the east of the district the lower sediments of the Calciferous Sandstone Measures and the lower Garleton Hills Volcanic Rocks have been folded into a series of shallow anticlines and synclines with approximately NE–SW axes. The sediments form low, drift-covered ground in the cores of the Balgone, Crauchie and Traprain anticlines, while the volcanic rocks outcrop in hills in the cores of the intervening faulted Whitekirk and Prestonkirk synclines. In the west the upper formations dip gently under the Lower Limestone Group. From Gullane to Bolton the general dip is westwards, while from East Saltoun to Crichton the dip is to the north-west. Although drift-cover obscures most of the structure, there is evidence of gentle folding as shown by the outcrop of the overlying limestones at East Saltoun, by the outlier of the Lower Longcraig Limestone at Lennoxlove and by the Devono-Carboniferous inliers at Humbie and Costerton. Likewise minor faulting seen along the coast and affecting the overlying limestone outcrops may be widely present.

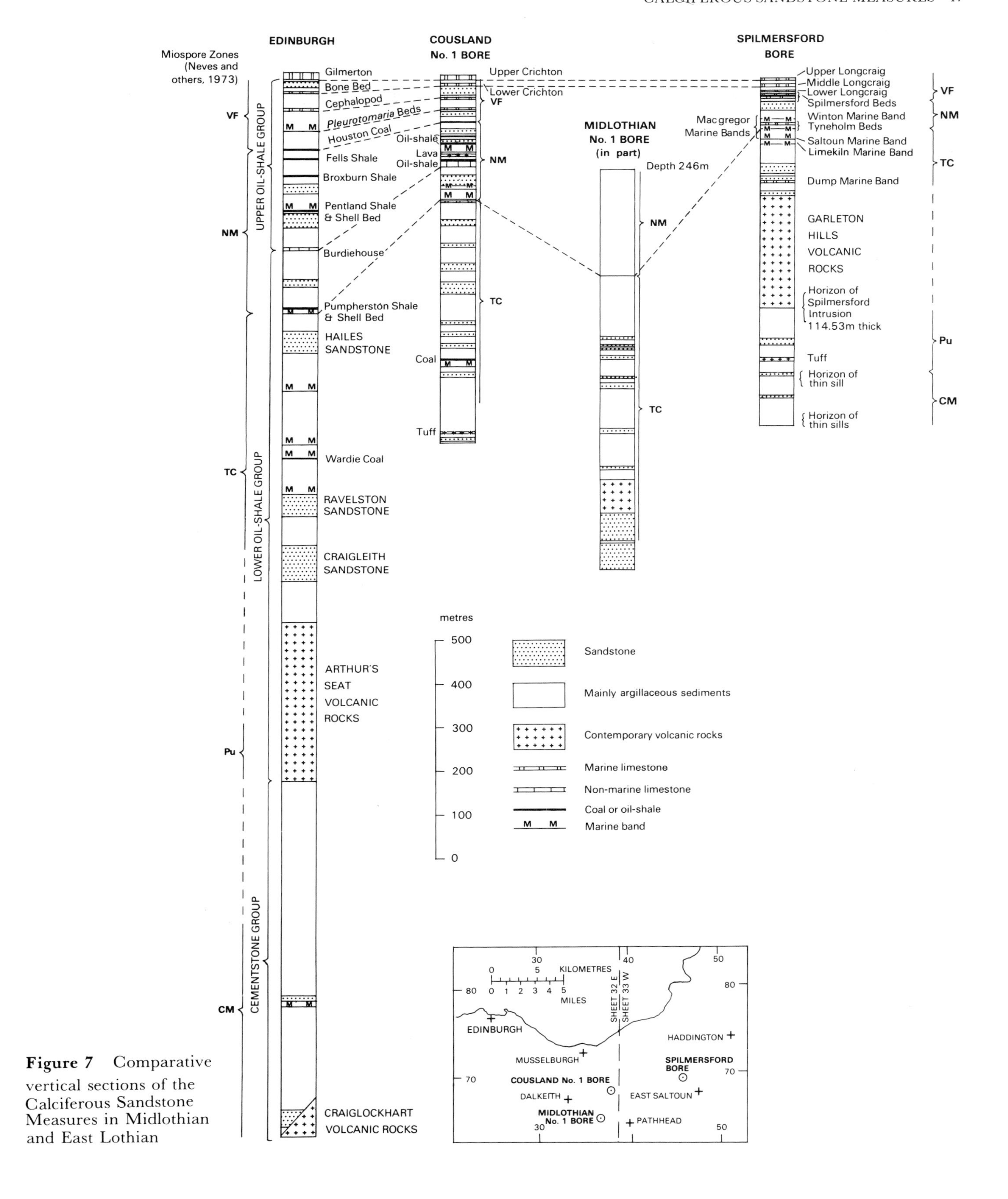

Figure 7 Comparative vertical sections of the Calciferous Sandstone Measures in Midlothian and East Lothian

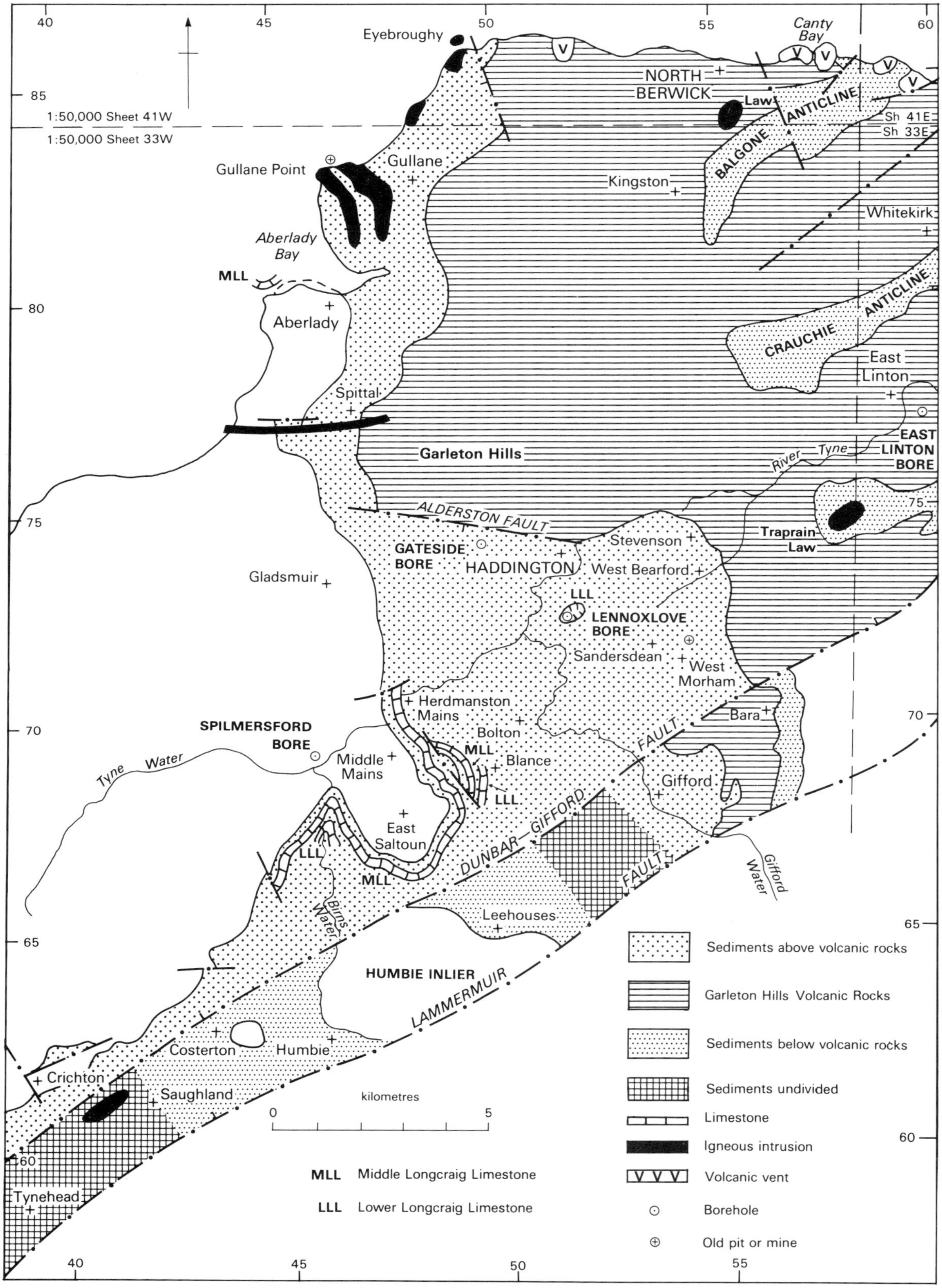

Sediments above volcanic rocks
Garleton Hills Volcanic Rocks
Sediments below volcanic rocks
Sediments undivided
Limestone
Igneous intrusion
Volcanic vent
Borehole
Old pit or mine

MLL Middle Longcraig Limestone
LLL Lower Longcraig Limestone

Figure 8 Generalised map of the Calciferous Sandstone Measures

DETAILS

Strata below Garleton Hills Volcanic Rocks

Spilmersford – East Linton

Sections through these strata were proved in the Spilmersford Borehole (Davies, 1974) and in the East Linton Borehole which was sited just east of the district boundary. The Garleton Hills Volcanic Rocks were passed through in both boreholes. In the former the base of the Carboniferous sequence appeared faulted against Upper Old Red Sandstone strata while the base was not reached at East Linton. The sediments were at least 266 m thick at Spilmersford, and over 346 m at East Linton. A thin development of pyroclastic rocks occurs in both sequences. Tentative correlation between the boreholes (Figure 9) is based on lithology as the non-marine faunas are not statigraphically diagnostic. If correlation is made of the brief pyroclastic episode in the middle of the two sequences, the strata below and above have comparable lithology and thickness.

The sediments below the pyroclastic rocks consist mainly of thick mudstones, silty mudstones and siltstones, variously grey, dark grey, green, red-brown and purple in colour. Cementstone nodules and cementstone bands up to 0.75 m thick are common at certain horizons. The sandstones are generally only a metre or two thick, fine- to medium-grained and white, grey, buff, reddish brown or purple in colour. One or two coal positions are poorly developed. Several horizons have a non-marine bivalve fauna, representing the *Modiolus*-phase of non-marine or quasi-marine conditions. No definite marine conditions were developed (Wilson *in* Davies, 1974, p. 12). The fauna at East Linton is similar, although less varied. In the deltaic sediments plant remains have been recorded.

In the pyroclastic episode at East Linton, purple and greenish tuffaceous sediments, tuffs and agglomerates are interbedded with normal sediments in which estheriids were found at intervals throughout 40 m of strata whereas at Spilmersford only 5 m of purple tuffs and agglomerates are present.

Sedimentation of the *Modiolus*-phase continued above the pyroclastics. Thick mudstones and siltstones with cementstone bands up to 0.6 m thick were developed and contain bands with a non-marine fauna. The lowest non-marine bands, some 8 to 10 m above the highest tuff band, at both Spilmersford and East Linton, contain almost identical faunas of the bivalves *Lithophaga* cf. *lingualis*, *Modiolus latus*, *Naiadites*?, estheriids and ostracods and are tentatively correlated. In both boreholes this band is overlain by a further 25 m of mudstone with cementstone bands, and succeeded by 15 m of grey channel-sandstone.

In both areas, strata in the upper 70 m or more are reddened. The mudstones and siltstones are stained red, red-purple and green-purple with green reduction spots and any sedimentary and fossil details are almost completely destroyed. Thick channel-sandstones, stained purple or bleached white due to secondary loss of the purple staining, are fine- to medium-grained, and show ripple bedding and large-scale contorted bedding. Passage from reddened sediments to tuff with thick cementstone bands is very gradual at East Linton, but clear-cut at Spilmersford.

Balgone Anticline

This most northerly of three anticlinal folds has sedimentary strata cropping out in the core from near Kingston to the coast at Canty Bay. On the shore there are faulted exposures of the massive white, brown and red, medium- to coarse-grained, cross-bedded Canty Bay Sandstone, showing carious-weathering features. Associated with the sandstones are reddened mudstones, siltstones and fine-grained sandstones (McAdam *in* Craig and Duff, 1975, p. 85). These strata are comparable with the highest reddened strata at

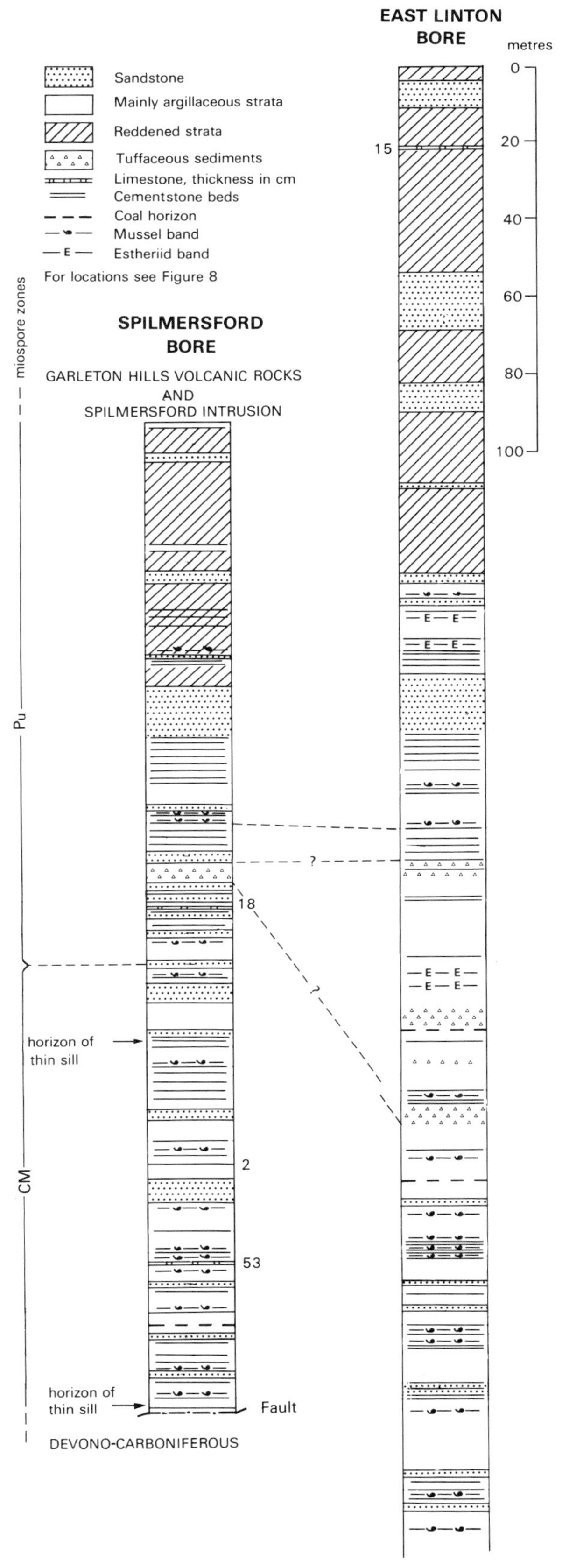

Figure 9 Comparative vertical sections of the lower Calciferous Sandstone Measures

Spilmersford and East Linton. Inland the sediments are largely obscured by drift, and the few exposures recorded in earlier surveys are no longer visible.

Crauchie Anticline

The outcrop area of the strata below the volcanic rocks is entirely drift-covered. One borehole record confirms the presence of 'marly' sandstone under 2.6 m of drift.

Traprain Anticline

Around Traprain Law the strata have been tilted at high angles, up to 80°, by intrusion of the phonolite laccolith. Just north of the Law, steeply dipping, dark red, micaceous sandstone was recorded from an old quarry, now obscured. Further to the north-west other outcrops showed greenish marl with sandy bands, green mudstone with cementstone nodules and bands of calcareous sandstone, and interbedded blue mudstone and yellow marl. These beds lie high in the sequence.

Bara area

Between the Dunbar–Gifford Fault and the Lammermuir Fault at Bara, there is an area of strata which lies between the outcrop of the Devono-Carboniferous and a thin development of the Garleton Hills Volcanic Rocks. The ground is drift-covered and little is known about the rocks.

Leehouses area

Strata dipping to the north-east lie on the Devono-Carboniferous rocks of the Humbie Inlier, between the Dunbar–Gifford and Lammermuir faults at Leehouses. The area is drift-covered and virtually nothing is known of the rocks. Their extent to the north-east is uncertain, as no outcrop of the overlying Garleton Hills Volcanic Rocks has been recorded. It is not known whether the outcrop is obscured by drift, or faulted out, or if the volcanic rocks are not developed in this area.

Humbie–Costerton area

Outcrops of the sequence below the volcanic rocks are present in the gently folded strata on the west side of the Humbie Inlier and around the Costerton Inlier of Devono-Carboniferous rocks. The only exposure within the district of the local base of the Carboniferous occurs in strata which dip to the SW at 8° to 12° in the Johnstounburn Water, 250 m E of Humbie [4615 6274]. The boundary is taken at the base of a 38-cm thick bed of decomposed red-brown and red-purple tuff and agglomerate. The overlying sediments are beds of red, purple, brown and grey, fine- to medium-grained sandstone, interbedded with soft variegated red, purple, pale green and brown silty mudstone, and mudstone containing decomposed dark red concretions and beds of soft argillaceous sandstone. On the west side of the Costerton Inlier is an exposure in the east bank of the Costerton Water, east-north-east of Costerton, of a thin red ironstone band containing Carboniferous plant remains and fish scales. The Carboniferous strata have faulted contacts to both north and south with the Devono-Carboniferous red, brown and purple sandstones. Because of the drift cover and the probable lack of the Garleton Hills Volcanic Rocks, the extent of the outcrop to the south-west towards Tynehead in unknown.

In the south-west border of the district around Tynehead, be tween the Dunbar–Gifford Fault and the Lammermuir Fault, are Calciferous Sandstone Measures strata of uncertain affinity. In the upper part of Maggie Bowies Glen, north-west of Tynehead the exposures show red, purple and brown, fine- and medium-grained sandstone, some beds being hard and siliceous, others soft and argillaceous. The beds dip at 5° to 10° to the NW, but in places are almost flat. From their colour and lithology they are most likely to be below the volcanic rocks.

Strata from top of Garleton Hills Volcanic Rocks to base of Macgregor Marine Bands

Spilmersford–Samuelston

The only complete section of this succession was recorded in the Spilmersford Borehole where it measured 135 m. The upper 65 m of the sequence is thought to have been encountered in the much earlier Samuelston No. 1 Borehole.

Above a thin basal volcanic conglomerate, lying on top of the Garleton Hills Volcanic Rocks, the sediments take on a rhythmic character. In the lower part of the formation there are four thick deltaic sandstones, fine- to medium-grained with coarse bands, and pale grey, buff or pale purple in colour. Otherwise the strata are mainly grey and dark grey siltstone and mudstones with plant remains. The cycles have coal positions with root-beds developed and thin coals up to 0.35 m thick have been recorded. Several cycles contain mudstones and limestones with sparse non-marine faunas. Marine conditions prevailed at two horizons, the Dump Marine Band which is 40 m above the volcanic rocks and the Limekiln Marine Band, some 90 m higher in the sequence. The Dump Marine Band is a brownish grey, sandy limestone 38 m thick, and the Limekiln Marine Band is a silty mudstone 1 m thick. Both bands contain a sparse fauna mainly of marine molluscs.

Spittal–Gullane

In the north of the district, where the Garleton Hills Volcanic Rocks reach their maximum thickness, deposition of the overlying sediments began later than in the area to the south where the volcanic sequence is thinner. North of the Alderston Fault around Spittal, the outcrop width between the volcanic rocks and the base of the Lower Limestone Group is very narrow. This was taken by Barrow (*in* Clough and others, 1910, p. 83) as evidence for a local unconformity at the top of the volcanic rocks.

Eybroughy to Gullane shore

Several discontinuous exposures interrupted by intrusions and faults, occurring along the coast from Weaklaw Rocks, opposite Eyebroughy, to Gullane Point (Figure 10), probably belong mainly to a thicker development of the sediments than is present to the south. The sediments are dominantly non-marine sandstones, siltstones and mudstones with plant remains and ironstone nodules. The Cheese Bay shrimp band and thin oil-shales are developed at one locality [492 857]. Tuffaceous and dolomitic sediments lie on the top of the Garleton Hills Volcanic Rocks.

East of the Archerfield Fault at Weaklaw Rocks there are gently folded cementstone-facies strata with a general dip to the north-west (Plate 17). Thin red-weathering grey cementstones are interbedded with grey siltstones and mudstones with plant fragments and rare shells. West of the fault, quite different sediments are folded into an asymmetric syncline surrounded by a ring fracture suggesting collapse and resembling the east Fife intrusion breccias (Francis, 1960, p. 44). Hard dolomitic bands alternate with softer tuffaceous bands (McAdam *in* Craig and Duff, 1975, p. 90).

Opposite Eyebroughy, outcrops of sedimentary rocks are interrupted by exposures of thin irregular basic sills. The thinly bedded, dolomitic mudstones, siltstones and sandstones with tuffaceous bands, black mudstones and hard brownish-grey limestone bands are greatly disturbed, veined and brecciated by the intrusions (Day, 1916b).

The inlet to the south, referred to in geological literature as Cheese Bay (Figure 10), has the following section of gently folded strata noted for rich faunas of arthropods and fish:

	Thickness m
Cementstone, hard, ferruginous; numerous small crustacea, mainly *Tealliocaris loudonensis*, fish remains	0.15
Mudstone, in part oil-shale, dark, fissile	0.5
Mudstone, dark purplish grey, paler bands; numerous fish	0.6

(Traquair, 1907; Peach, 1908; Clough and others, 1910, pp. 84, 207, 215–217; Hesselbo and Trewin, 1984).

In the south part of the bay, black and grey mudstones with plant remains and hard grey, yellow-green weathering, limy mudstones with fish and plant remains are seen passing under the small sill forming the headland. In the next inlet south of the fault which truncates this sill, westerly-dipping, mainly grey, ochreous-weathering, very fine-grained dolomitic sediments are exposed. Although the next 500 m of foreshore is occupied by the main outcrop of the Cheese Bay Sill, the sediments overlying the sill are exposed. These include soft red and grey tuffaceous rocks, passing up into hard, ochreous-weathering dolomite, disturbed and brecciated by net-veining of the red-stained tuff (Day, 1916a).

On the shore north-west of West Links the south-west dipping rocks exposed are mainly hard, coarse, yellow-brown-weathering dolomitic conglomerates containing fragments and blocks of grey, green or yellow cementstone, fine-grained very quartzitic sandstone, green and purple striped tuffs and decomposed lavas. On the south side of the same bay, north-east of the Black Rocks Sill, soft, pale green, tuffaceous rocks dip gently south. These are overlain by hard, massive, well-bedded, brown-weathering dolomite locally mottled dark red, buff and green, with beds of volcanic detritus containing angular basalt lava fragments up to 8 mm long.

The Bleaching Rocks are exposures of white and grey, medium- to fine-grained sandstones with silty bands, dipping to the south-west at 10° to 27°. These sandstones are thickly bedded in places and consequently have been exploited in three inland quarries. In Laird's Quarry, seatclay and thin coals were exposed, interbedded with the purple and grey sandstones. White sandstones with beds of mudstone and two thin coal seams were seen in the infilled Whim Park Quarry (Barrow *in* Clough and others, 1910, p. 85). Gullane

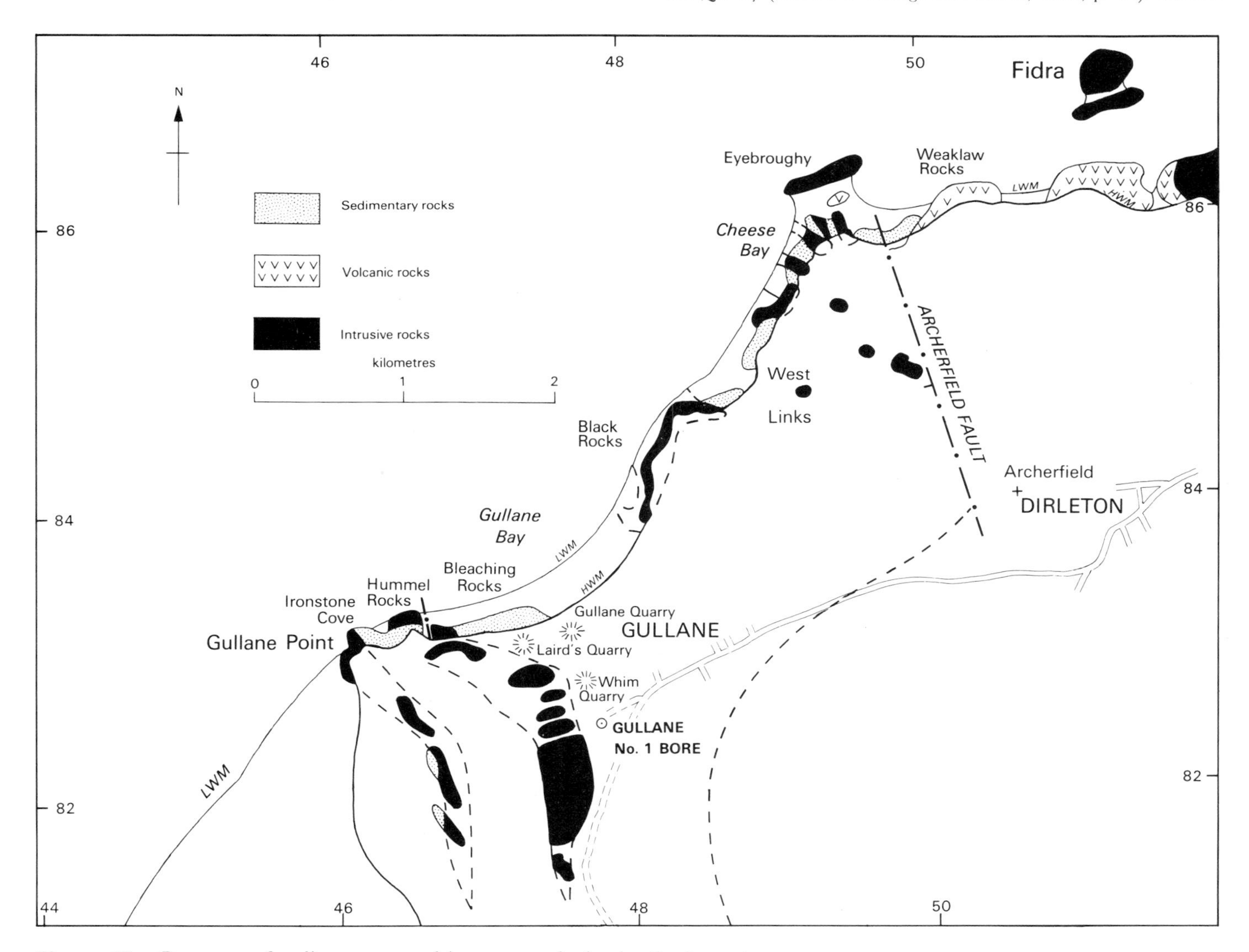

Figure 10 Outcrops of sedimentary and igneous rocks in the Eyebroughy –Gullane area

Links No. 1 Borehole (1894) proved over 60 m of strata below those seen in Whim Park Quarry and approximately equivalent to the rocks just described from the coast. Thick beds of white sandstone alternate with mudstones and ?fireclays, some with ironstone nodules. A 0.4 m thick, impure limestone in the upper part, and two thin bands of carbonaceous mudstone are possibly equivalent to the beds in Cheese Bay.

West of the headland formed by the teschenite sill at Hummel Rocks are folded and faulted exposures in a little bay called Ironstone Cove (Figure 10). Grey-white, ochreous-weathering sandstones with beds of conglomerate alternate with seatclays and mudstones containing ironstone nodules. It was mining of the nodules which gave the bay its name. The strata dip generally south-west under a higher teschenite sill forming Gullane Point.

Haddington

South of the Alderston Fault an outcrop area of the succession above the volcanic rocks underlies the town of Haddington. The strata were encountered in several old boreholes which tap water from thick sandstones in the sequence. A recent borehole at Gateside proved thick sandstones in the lower part of the succession. The Dump Marine Band was not recorded under the thick sandstone at Gateside but marine and non-marine fossils are present in the limestones below. Volcanic rocks immediately underlie these limestones. A coal seam, 0.2 to 0.45 m thick, was formerly exposed at the St Laurence Burn, west of Haddington.

Sandersdean

Exposure is poor along the outcrop of the sequence from Stevenson through Sandersdean to the Dunbar–Gifford Fault. Of note is a coal seam, probably high in the succession, which was recorded as 0.6 m thick in the Dean near Sandersdean, and was worked in shallow pits just north of West Morham. This is possibly on the horizon of the coal seen near Haddington (above) and also a 0.2-m coal formerly worked on the north bank of the Tyne, south-west of Haddington, upstream from Samuelston.

Gifford

Little is known about the strata which overlie the Garleton Hills Volcanic Rocks between the Dunbar–Gifford and Lammermuir Faults around Gifford. Exposures in the Gifford Water south-east of the village show a massive white and red sandstone overlying the volcanic rocks. Dark mudstone and thin coals occur in strata at higher horizons.

Strata from Macgregor Marine Bands to base of Upper Longcraig Limestone

A complete section of this sequence, 133 m thick, was recorded in the Spilmersford Borehole. At Lennoxlove, near Haddington, an old borehole proved all but the top part of the succession and indicated a similar thickness. The upper limestone-bearing beds are exposed on the shore at Aberlady, but inland only short sections of strata are exposed in streams.

The sediments, as seen at Spilmersford, have a marked rhythmic character. In the middle of the sequence thick deltaic sandstones and thick mudstone and siltstone bands containing plants are developed. Numerous seatearths are present, but little or no coal is developed. Seven marine horizons were recognised in the sequence at Spilmersford. The three lowest ones, in the basal 50 m, are the Saltoun Marine Band, the Tyneholm Beds and the Winton Marine Bands, which together form the type succession of the Macgregor Marine Bands (Wilson, 1974). These are the lowest beds in the Carboniferous of the district which contain a rich marine fauna, and they can be traced over much of south-east Scotland. Each marine horizon has a thick development of marine mudstones and thin limestones and contains a varied, rich, marine fauna. The Sandersdean Limestone of the Haddington area has been correlated with the Macgregor Marine Bands (Wilson, 1974). The other four marine horizons occur in the top 40 m of the sequence. These are the Spilmersford Beds, the Lower Longcraig Limestone, the Middle Longcraig Limestone, and the marine mudstones below the Upper Longcraig Limestone.

Macgregor Marine Bands

The Macgregor Marine Bands was the term proposed by Wilson (1974) to denote the first occurrence of rich marine faunas in the Lower Carboniferous of south-east Scotland. The type sequence was taken in the succession proved by the Spilmersford Borehole (Davies, 1974) and includes the Saltoun Marine Band, the Tyneholm Beds and the Winton Marine Band in ascending order. The three separate marine horizons, spread over about 30 m of an otherwise non-marine sequence, at the type locality is the maximum number recorded. More commonly, only one or two bands are present and the sequence is highly variable over relatively short distances (Wilson, 1974, p. 42).

In the Spilmersford Borehole the Saltoun Marine Band consists of 6.4 m of silty mudstones, the Tyneholm Beds are composed of about 10 m of limestone and calcareous sandstone and the Winton Marine Band is about 1 m of mudstone and siltstone. All three horizons yielded rich and varied marine faunas.

The Sandersdean Limestone was first described by Bailey (*in* Clough and others, 1910, p. 56) from its type locality [5361 7156] south-east of Haddington. This marine horizon and several other isolated exposures of marine bands in the district were correlated on faunal and miospore evidence by Wilson (1974, p. 63).

Macgregor Marine Bands to Spilmersford Beds

In the Spilmersford Borehole, this part of the succession is about 45 m thick and may thicken towards Lennoxlove. The notable features of the sequence are very thick sandstones, grey-white to pale or deep purple in colour, fine-, medium- and coarse-grained, with pebbly bands, and showing cross-bedding, slumped bedding and other deltaic features. Interbedded with these are grey, purple and yellowish green mottled mudstones and siltstones.

In numerous old sandstone quarries to the east and south-east of Haddington these sandstones were worked, but it is not known exactly which sandstones were quarried at the individual sites. Yellow and white sandstones low in the formation were exploited in quarries at Rentonhall [544 720], just north-east of Sandersdean. Possibly the same sandstone horizons are exposed in the banks of the Bearford Burn [543 725] upstream from West Bearford, and were worked in a quarry at Woodend [536 725], 1 km S of Sandersdean, and in two quarries due east of Haddington (Bailey *in* Clough and others, 1910, p. 55). A grey-white, red-stained sandstone [515 727] formerly quarried 800 m N of Lennoxlove is equated with the thick sandstone lying below the Lower Longcraig Limestone in that borehole.

Spilmersford Beds

In the Spilmersford Borehole (Davies, 1974, p. 5) these beds are 10.72 m thick, composed of 7.44 m of alternating limestones and mudstones, containing marine fossils, with a 3.28-m thick inter calation of grey sandstone.

The faunal assemblage bears a close relationship to that recorded from the limestones and shales at Ardross, near St Monance on the Fife coast, which also lie near the top of the Calciferous Sandstone Measures (Wilson *in* Davies, 1974, p. 11). The limestones and mudstones of the Spilmersford Beds have not been recognised at outcrop in the district. Some 3.53 m of rooty beds, with sphaerosiderite common in the lower half, overlie the Spilmersford Beds in the borehole.

Lower Longcraig Limestone

At Spilmersford this limestone was represented by 1.22 m of grey, shelly and crinoidal limestone containing a rich marine fauna with abundant corals, brachiopods and molluscs.

In the East Saltoun area, some 2 km to the south-east, the Lower Longcraig Limestone has been proved to be around 1 m thick in boreholes. The bed is partly exposed in the Blance Burn [489 687] near Blance, north-east of East Saltoun, where 0.46 m of dark-grey limestone with mudstone bands containing corals, brachiopods and crinoid fragments is present. A fine-grained grey limestone exposed in the Birns Water [460 676], west of the village may be the same limestone. An outlier of the Lower Longcraig Limestone at Lennoxlove was worked in the Belvidere Quarry [514 722], site of the Lennoxlove Borehole. About 2 m of flaggy and argillaceous crinoidal and shelly limestone with calcareous mudstone partings is exposed. A bed of crinoidal limestone, 0.9 m thick, formerly exposed in a quarry east of Aberlady may be from this horizon.

In the Spilmersford Borehole the interval between the Lower Longcraig Limestone and the Middle Longcraig Limestone is 10.8 m, consisting of the 1.7 m of calcareous marine shales overlying the lower limestone, 5 m of grey silty mudstone with irony nodules and plants, a grey siltstone and a rooty sandstone with a coal position immediately underlying the higher limestone. At Aberlady the interval is reduced to around 4 m and is sandier, but the coal position is still present.

Middle Longcraig Limestone

Throughout the district the Middle Longcraig Limestone is about 4 m thick and is characterised by abundant corals, in places being composed largely of the colonial coral *Lithostrotion junceum*. Locally the overlying marine and deltaic phases are absent and the upper part of the limestone has been decomposed and altered to pale mudstone with limestone nodules by the bleaching action of a later cover of vegetation which is evidenced by an overlying seatclay and thin coal.

The Middle Longcraig Limestone is well exposed on the Aberlady shore, north-east of Craigielaw Point (Figure 13). Here the bed is about 4 m thick, white, grey or yellow-brown in colour, composed largely of *Lithostrotion* with large numbers of the brachiopods *Composita* and *Pleuropugnoides*. At Spilmersford the limestone was 4.51 m thick, grey with pale grey nodules and yielded a rich marine fauna.

In the East Saltoun area, to the south-east, boreholes have proved that the limestone ranges in thickness from 1.5 to 4.85 m. The outcrop of the Middle Longcraig Limestone has been mapped round the Saltoun Basin, and several sections recorded. To the north of East Saltoun, in the banks of the River Tyne below Herdmanston Mains are exposures of a cream-coloured nodular limestone with corals and brachiopods. To the east of the village 2 m of hard nodular, very fossiliferous limestone with corals, brachiopods and bivalves was formerly exposed in the old Saltoun Lime Works quarries at Middle Mains. The limestone was quarried on the banks of the Birns Water west of East Saltoun, where 3.65 m of grey, flaggy and massive, crinoidal and shelly limestone with corals in the upper part is seen. There are no exposures of the Middle Longcraig Limestone between the Birns Water and the western margin of the district.

Strata above the Middle Longcraig Limestone

The Middle Longcraig Limestone is generally overlain by some 5 m of non-marine siltstone and sandstone capped by a thin coal. This seam is 0.25 m thick in some boreholes around East Saltoun, though only 0.01 m thick at Spilmersford. At 0.46 m the seam was thick enough to be mined in a pit north-east of Crichton. The 4.72 m of dark grey, calcareous, silty mudstone with bands of impure limestone and shells overlying the thin coal at Spilmersford underlies the Upper Longcraig Limestone and is the uppermost bed of the Calciferous Sandstone Measures.

LOWER LIMESTONE GROUP

The Lower Limestone Group is the uppermost division of the Lower Carboniferous in the district. Its limits coincide approximately with those of the Upper Bollandian Stage (P_2) of the goniatite zonation (Currie, 1954, p. 534) and the group was placed in the Brigantian Stage by George and others (1976, p. 47). The base of the group is drawn at the base of the Upper Longcraig Limestone which has been correlated on lithological and faunal grounds with the Hurlet Limestone of the Glasgow area (Wilson, 1974, pp. 38–39), the long accepted base of the group in central Scotland. The top of the group is taken at the top of the Top Hosie Limestone, a well-known horizon over much of the Midland Valley. In the present district the limestone itself is not always present and the boundary is then drawn at the base of the marine mudstone which is associated with the limestone in other areas.

The total thickness of the Lower Limestone Group was proved to be 150 m in the Cousland No. 3 Borehole sited 3 km SW of Tranent just west of the western margin of the district. The thickness of the group probably decreases in the east and south-east across the district from this borehole (Goodlet, 1957, pp. 56, 58), and is estimated to be 80 m in the Pencaitland area. This decrease is due to thinning of beds rather than unconformity, as bed by bed correlation of the main horizons in the lower part of the group between East Lothian and Midlothian is well established (Figure 11). The lower limestones can be traced along the outcrop round the south side of the East Lothian Coalfield. In the upper part of the group there is little faunal evidence and correlation of the borehole sections available is tentative. However, it is likely that the Lower Vexhim Limestone, Upper Vexhim Limestone and Bilston Burn Limestone are all developed in at least part of the district, as is the Top Hosie Limestone. Correlation of the upper part of the Lower Limestone Group into the Dunbar area (Figure 11) is less certain. The Barness East Limestone is taken as equivalent to the Top Hosie Limestone, but it may in fact correlate with one of the lower limestones.

The outcrop of the Lower Limestone Group (Figure 12) occurs on the coast between Aberlady and Longniddry. Inland the outcrop extends south-south-west through Gladsmuir to East Saltoun, and then to the south-west to Pathhead and Crichton on the western margin of the district. The general inclination of the beds from the coast to East Saltoun is towards the west or west-south-west, while from East Saltoun to the south-west margin of the district, the strata dip in a north-westerly direction under the Limestone Coal Group strata of the East Lothian Coalfield. The width of the outcrop varies considerably, being controlled largely by the dip of the strata and the local topography. Around

Gosford Bay and East Saltoun, for example, the outcrop broadens because of gentle basin structures. The upper beds of the group outcrop also in the core of the D'Arcy–Cousland Anticline on the west side of the Crossgatehall Fault near Bankhead west of Tranent.

The lower part of the Lower Limestone Group sequence is well exposed on the shore west of Aberlady (Plates 4,5). The strata between the Skateraw Limestone and the top of the group are very poorly exposed along the south shore of Gosford Bay, south-west of Ferny Ness. Inland the beds are largely obscured by superficial deposits, with only rare natural exposures, and are known mainly from boreholes. The Upper Longcraig Limestone and the Skateraw Limestone were formerly exposed in numerous quarries, but most of these excavations are now filled in or overgrown. Details of some of these quarry sections were given in Clough (*in*

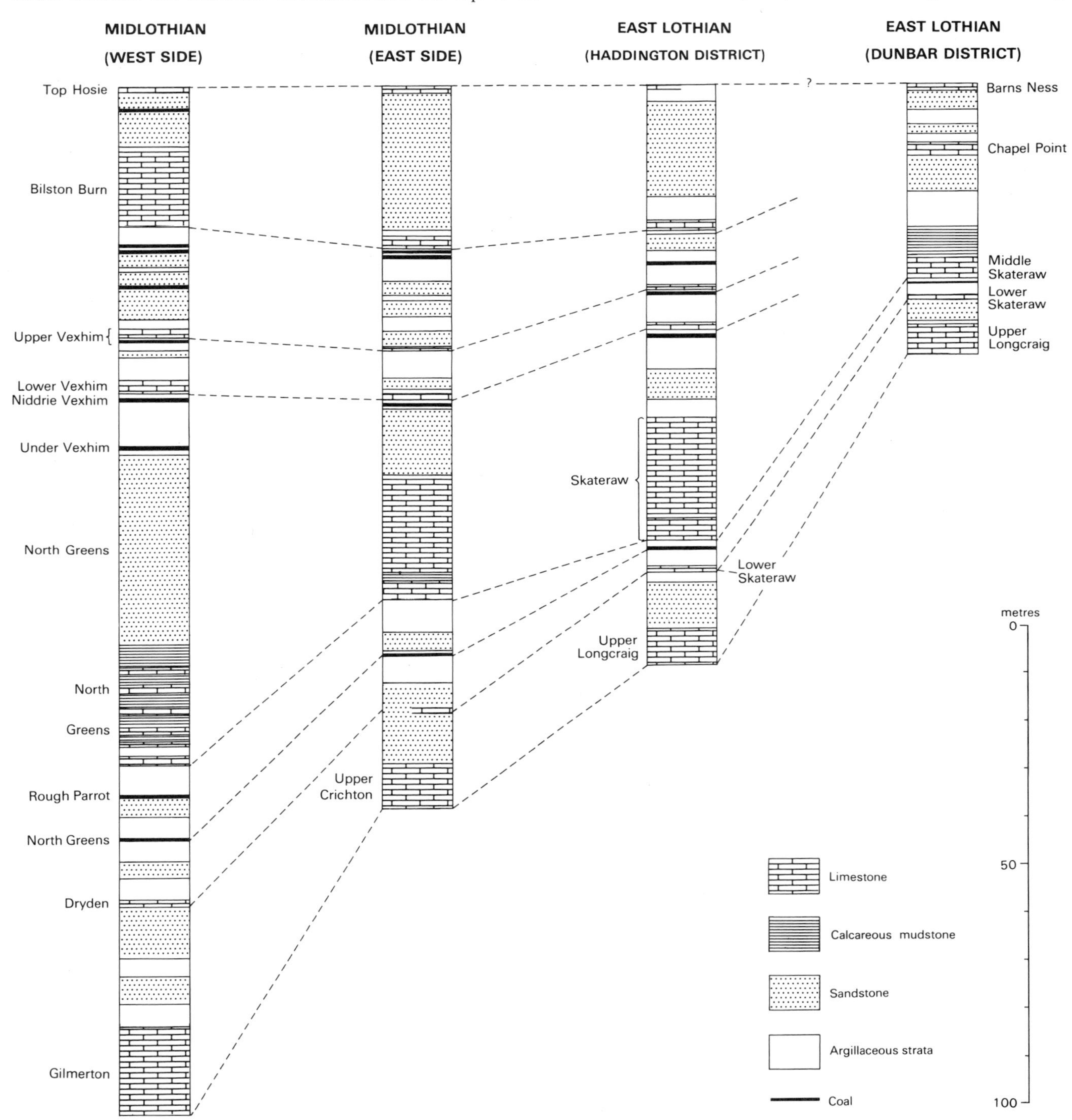

Figure 11 Correlation of the Lower Limestone Group in Midlothian and East Lothian

Clough and others, 1910, pp. 138–145) and by Robertson and others (1945). Lists of fossils found in these quarries and in natural exposures have been published by Lee *in* Clough and others (1910, pp. 209–212) and by Wilson (1974, pp. 53–65). The complete succession of the Lower Limestone Group is not seen in natural sections in the district and has not been proved in any single borehole. A general section has been drawn up (Figure 11).

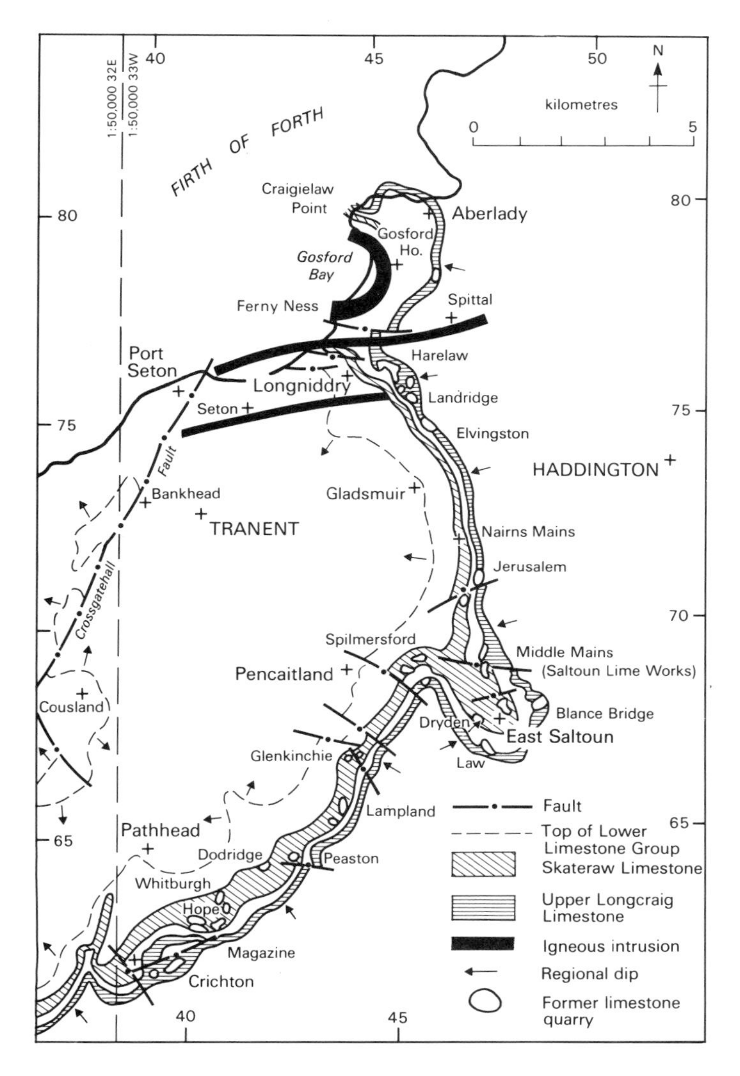

Figure 12 Generalised map of Lower Limestone Group outcrops

Structure

The Lower Limestone Group strata have a regional dip to the west or north-west. In general the strata dip at up to 5°, but in some places the beds are flat-lying and in others dips of up to 16° have been recorded. Broad synclines affect strata of the group at Gosford Bay, where the axis plunges gently westwards, and in the East Saltoun area with a gentle plunge to the north-north-east. At several localities gentle minor folds are superimposed on the larger structures. The most notable of these are exposed on the Kilspindie shore between Craigielaw Point and Aberlady Point where a gentle anticline and syncline, both with axis plunging to the south-east, lie on the north limb of the Gosford Bay syncline. Several minor faults cut the Lower Limestone Group strata and displace the outcrop of the limestones.

Lithology

The Lower Limestone Group succession is essentially a cyclic development of marine limestone, mudstone, sandstone, seatclay or seatearth and coal deposited partly in a marine and partly in a deltaic environment. The recurring cycles indicate subsidence of a land area below a shallow sea, the formation of a marine limestone, the subsequent shoaling of the sea by the accumulation of mud, silt, and sand, and the formation of a fresh land surface. The coal seams represent vegetation which grew on the swampy land surfaces, and which were subsequently buried by renewed subsidence. The cycles are frequently broken or incomplete. The rich marine faunas present in the uppermost beds of the Calciferous Sandstone Measures continue into the lower part of the Lower Limestone Group, and mark the period when marine influences reached their maximum in Carboniferous times within the district.

In the district, the succession is composed dominantly of limestones, mudstones and sandstones. The coal seams, seatclays and seatearths are thin. In the lower part of the group more than half the sequence is composed of limestone and calcareous mudstone. Two thick limestones are present, the Upper Longcraig Limestone, and the Skateraw Limestone. The available information indicates that the limestones in the upper part of the Group are fairly thin. Thick beds of sandstone occur at several horizons.

DETAILS

Upper Longcraig Limestone to Skateraw Limestone

The only good natural section through the lower part of the group is seen on the Kilspindie shore between Craigielaw Point and Aberlady Point on the south side of Aberlady Bay, west of Aberlady (Figure 13). Only the lower 3.6 m of the Skateraw Limestone is exposed, but the complete sequence is present below including the Lower Skateraw Limestone, the Upper Longcraig Limestone and the intervening strata. The rocks on the Kilspindie shore are gently folded with minor faulting. The crest of an anticline with a NW–SE trend crosses the shore 250 m NE of Craigielaw Point, and the axis of the complementary syncline on a parallel trend lies 200 m SW of Aberlady Point. As a result of the folding, the Upper Longcraig Limestone outcrops repeatedly at Craigie Point, Garlick Rocks (Plate 5) and Aberlady Point, with the overlying strata also repeated.

The general sequence of beds exposed along the Kilspindie shore is:

	Thickness m
Skateraw Limestone, grey, hard, thickly-bedded, crinoidal	3.6 +
Strata, including a thin coal, seatearth, pyritic silty mudstone with irony nodules, micaceous sandy siltstone (exposure poor)	6.1

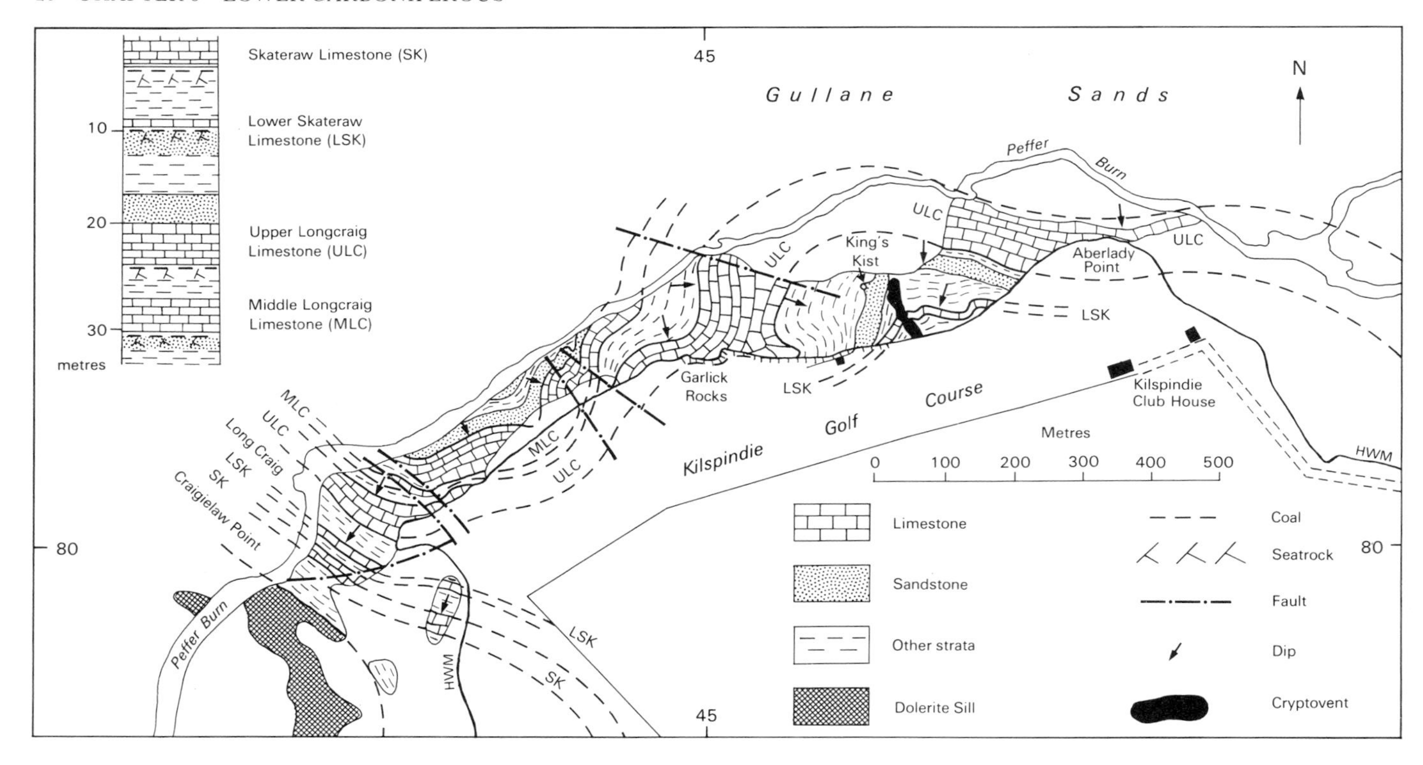

Figure 13 Map of the solid geology on the shore at Kilspindie

	Thickness m
LOWER SKATERAW LIMESTONE, grey, hard, dolomitised, crinoidal, *Gigantoproductus* cf. *giganteus*	0.6 to 1.2
Coal and coaly mudstone	0.08
Seatearth	1.0
Sandstone, yellow-weathering, medium-grained, cross-bedded, ripple-bedded, sharp base	2.1
Silty sandstone to silty mudstone, grey, thinly-bedded, ripple-bedded, bioturbated	3.5
Sandstone, pale grey, flaggy	0.45
Mudstone, grey and black, soft, fissile	0.9
Sandstone, yellow, fine- to medium-grained, cross-bedded, contorted bedding, ripple-marked, worm tracks, pyritic patches	1.8
Silty mudstone, dark grey, calcareous, few ironstone nodules	0.5
UPPER LONGCRAIG LIMESTONE, grey, reddened upper surface; upper part coarse, flaggy, thin silty bands, crinoidal; lower part nodular, crinoidal, shelly with *Lithostrotion junceum*	3.6 to 5.5

Upper Longcraig Limestone

The Upper Longcraig Limestone is normally 3 to 5 m thick. It is generally grey, nodular, and crinoidal, and contains a very rich and varied fauna of corals, bryozoa, brachiopods and molluscs (Wilson, 1974), most notably large colonies of the coral *Lithostrotion junceum.*

Some 3 km S of the Kilspindie shore section described above, two exposures of the Upper Longcraig Limestone, over 3.6 m of grey, dolomitic, crinoidal limestone, occur on either side of a small fault, on the foreshore between Ferny Ness and the Port Seton–Spittal quartz-dolerite dyke. Inland, crinoidal limestones worked in a quarry east of Gosford House and proved in a borehole 400 m N of the quarry are taken as the Upper Longcraig Limestone. From Longniddry to Elvingston, where the strata are gently folded and dipping towards the west and south-west at low angles, the Upper Longcraig Limestone was exploited in several overgrown or grassed quarries. In the largest of these at Harelaw Limeworks [4506 7633], the limestone was at least 4.6 m thick and overlain by 3.6 m of calcareous mudstone.

There are no sections or borehole records in the Upper Longcraig Limestone south from Elvingston for 5 km to the East Saltoun area, where the bed was quarried extensively round the syncline. The limestone is known from numerous boreholes as well as records from the quarries where as much as 5.8 m of limestone was exposed.

Just to the north, in the Spilmersford Borehole, put down in Spilmersford Quarry, the limestone was 2.89 m thick, grey, fine-grained, crinoidal, with a 0.1-m band of *Lithostrotion* and a rich fauna of brachiopods (Davies, 1974).

Little is known about the Upper Longcraig Limestone from the East Saltoun area along the outcrop south-west to Crichton. A nodular crinoidal limestone representing the top of this bed crops out in the Kinchie Burn [4459 6684], 120 m downstream from the Glenkinchie Distillery.

In the Crichton area the outcrop is much faulted. In one borehole the Upper Longcraig Limestone, known locally as the Upper Crichton Limestone, is over 10 m thick in leaves. Over 6 m of hard concretionary limestone with thin mudstone laminae was seen in quarries [394 617] 750 m SE of Crichton, though neither the top nor the base was exposed. The limestone is very fossiliferous with corals, brachiopods and crinoid debris.

The interval between the Upper Longcraig and Lower Skateraw limestones is very variable, even over short distances. Boreholes in the East Saltoun area proved this interval ranges from 1.2 to 11 m, but is generally about 6.5 m. At East Saltoun the intervening strata consist mainly of red-stained, fine- to medium-grained sandstone. Pale yellow and white sandstone, over 2.4 m thick, underlies the Lower Skateraw Limestone in the northerly quarry at Jerusalem [468 705].

Plate 4 Upper Longcraig Limestone, Lower Limestone Group, Kilspindie, Aberlady (D3540)

Lower Skateraw Limestone

This bed is usually 0.6 to 1.2 m of grey, dolomitised, crinoidal limestone. The Lower Skateraw Limestone does not have such a rich fauna as either the Upper Longcraig Limestone below, or the overlying Skateraw Limestone, but it is distinguished by containing the maximum development of *Gigantoproductus* cf. *giganteus* in the succession.

The limestone crops out on the Longniddry shore 600 m S of Ferny Ness on the north side of the quartz-dolerite dyke where it is a hard, grey, dolomitic limestone, 1.2 m thick.

Apart from a record of a 0.9-m thick limestone in the northerly quarry at Jerusalem, nothing is known of the Lower Skateraw Limestone between the coast and East Saltoun, where it has been recorded in several boreholes. The limestone is generally 1.4 m thick. The lower 0.9 m of the bed, fine-grained, crinoidal with *Lithostrotion* towards the base, was exposed at the north end of Law Quarry, south-west of the village. In the Spilmersford Borehole the limestone was 1.42 m thick and contained bands of *Lithostrotion.*

Along the southern outcrop little is known about the Lower Skateraw Limestone. Two outcrops, separated by a NW-trending fault, occur in the Kinchie Burn. A one-metre bed of hard grey limestone with scattered crinoidal and shelly debris was exposed in an excavation 50 m downstream of the Glenkinchie Distillery and also 45 m upstream of the bridge west of the distillery. Between Hope and Crichton the Lower Skateraw Limestone has been proved in boreholes to be 1.5 m of grey, crinoidal limestone. At Crichton the limestone is locally absent, its position occupied by sandstone.

The interval between the Lower Skateraw Limestone and the Skateraw Limestone varies from 5 m on the Aberlady shore to as much as 10 m in the East Saltoun area; it was 6.33 m in the Spilmersford Borehole. Two coals, locally present in this interval in the west of the district, probably represent the North Greens and Rough Parrot coals of Midlothian. At Crichton the probable equivalent of the North Greens Coal is 0.53 m thick and lies 5.5 m below the Skateraw Limestone in one borehole; in a nearby borehole the seam is reduced to a carbonaceous mudstone, but has the typical *Lingula* band in the roof. Both seams are thin in boreholes in the East Saltoun area.

Skateraw Limestone

In both the Haddington and Dunbar districts, the beds described as the Middle Skateraw limestones and Upper Skateraw Limestone are part of the same marine cycle. Indeed in the former area the upper part of the thick marine development contains several beds of limestone. The complete sequence has been named the Skateraw Limestone. Although it is the thickest limestone in the Lower Limestone Group, only the lower part is pure enough to be worked for lime.

The Skateraw Limestone is the equivalent of the North Greens Limestone of Midlothian. At Cousland the limestone is over 25 m thick, but generally the Skateraw Limestone in East Lothian is 10 to 20 m thick. It occurs in beds of massive crinoidal limestone separated by bands of fossiliferous calcareous mudstone which contain the most varied fauna in the succession.

The limestone is well exposed and is 3.6 m thick at Aberlady but

Plate 5 Cross-bedded sandstone overlying laminated silty mudstone, above Upper Longcraig Limestone, Garlick Rocks, Aberlady (D3537)

its outcrops on the Longniddry shore are separated by faults and by a quartz-dolerite dyke. At the seaward end of the outcrop north of the dyke 0.75 m of dark grey, flaggy limestone is overlain by 1.9 m of grey crystalline limestone with crinoidal debris and a few thin silty laminae. The upper part of the limestone is truncated by a strike fault, and along the outcrop near H.W.M. it comes in contact with the quartz-dolerite dyke. About 140 m S of the dyke is exposed a 0.6 m bed of grey, crinoidal limestone.

Information on the Skateraw Limestone along the eastern outcrop is lacking from Longniddry south to Nairns Mains. A borehole there encountered 9.7 m of limestone beds separated by bands of calcareous mudstone, considered to be the lower part of the Skateraw Limestone. A further 1.4 km to the south, a section in this limestone was exposed in the southerly quarry at Jerusalem, where the strata dip very gently westwards and 6.55 m of limestone with a 0.5-m thick, intercalated bed of calcareous mudstone was recorded.

The full thickness of the Skateraw Limestone is not seen in the vicinity of East Saltoun. The many quarries in the area which exploited the basal good quality beds are mainly infilled or overgrown. Several borehole have proved up to 10 m of alternating beds of grey, crinoidal and shelly mudstone, calcareous mudstone and argillaceous limestone, overlying a 5.5-m basal bed of hard, grey crinoidal limestone. The limestone was worked extensively at several quarries along the outcrop including Spilmersford [453 688], Glenkinchie Distillery [445 669], Lampland [438 655], Peaston [428 644], Dodridge [420 642], Hope [402 629] and Magazine [406 629]. Details of some of these sections, now overgrown, were given by Clough (*in* Clough and others, 1910, pp. 144–145). The maximum recorded thickness of the limestone and associated calcareous mudstones was 21.5 m in boreholes north-east of Spilmersford.

Skateraw Limestone to Top Hosie Limestone

Detailed information about the succession in the upper part of the Lower Limestone Group is derived almost entirely from boreholes, some of which passed through only short stretches of strata. No borehole in the district passed through the entire sequence, the nearest such borehole lying beyond the western boundary at Cousland. Marine fossils have been recorded from thin limestones in a few boreholes in the district but the information is poor and the tentative correlations are made mainly on lithology.

Above the Skateraw Limestone in the Bankhead area, on the western margin west of Tranent, is a thick sandstone, correlated with the North Greens Sandstone in the Midlothian Coalfield

Plate 6 Carboniferous fossils

1 *Tealliocaris woodwardi* × 1, Calciferous Sandstone Measures, Gullane.

2 *Fenestella sp.*, *Stenodiscus haddingtonensis* × 1, Macgregor Marine Bands, Sandersdean Burn.

3 *Overtonia fimbriata* × 2, Upper Longcraig Limestone, Birns Water.

4 *Pugilis pugilis* × 1, Middle Longcraig Limestone, Kilspindie.

5 *Composita* cf. *ambigua* × 2, Middle Longcraig Limestone, Kilspindie.

6 *Pterinopectinella?* × 1.5, Tyneholm Beds, Spilmersford Borehole at 123 m.

7 *Pleuropugnoides* cf. *pleurodon* × 2, Middle Longcraig Limestone, Kilspindie.

8 *Naiadites* cf. *crassus* × 1, Macgregor Marine Bands, Sandersdean Burn.

(Tulloch and Walton, 1958, pp. 23–24). Some 15.4 m of medium- and coarse-grained sandstone with pebbly bands were present in a borehole near Bankhead and the sandstone is almost as thick at Cousland. Further east the sandstone is much thinner and the interval between the Skateraw and Lower Vexhim limestones is consequently much thinner. A thin coal developed between the sandstone and the Lower Vexhim Limestone in the west of the district is equivalent to the Under Vexhim Coal of Midlothian.

Lower Vexhim Limestone

The Lower Vexhim Limestone has only been recorded for certain within the district in a borehole near Bankhead where it was represented by a hard, brownish, crinoidal limestone, 1.73 m thick. A bed of grey, argillaceous, shelly and crinoidal limestone and calcareous mudstone 2.5 m thick, lying 15.25 m above the Skateraw Limestone in a borehole 2.5 km NW of East Saltoun, is tentatively correlated with the Lower Vexhim Limestone.

Upper Vexhim Limestone

Separated from the previous limestone by about 5 m of normally argillaceous strata, the Upper Vexhim Limestone has been proved in several boreholes in the north-west part of the district. In a borehole near Bankhead the limestone measured 0.71 m of hard, grey, crinoidal and shelly limestone lying 6.27 m above the Lower Vexhim Limestone. The bed was recorded in a few boreholes in the Tranent and Macmerry areas, the maximum thickness occurring being 1.27 m.

The measures between the Upper Vexhim Limestone and the Bilston Burn Limestone are over 20 m thick in the western edge of the district but thin eastwards to 8 m at Seton. They contain one or more thick sandstone beds.

Bilston Burn Limestone

About 12 m thick at the type locality in Midlothian (Tulloch and Walton, 1958, p. 25), the Bilston Burn Limestone is reduced to between 2 and 5 m near the western margin of the present district. In the district where the limestone is developed along most of the outcrop, it is normally 1 to 2 m thick. In a borehole near Bankhead it is correlated with a 0.43 m thick bed of hard, grey, red-stained, crinoidal limestone. In a borehole near Seton it was 2.82 m thick. This variation in thickness was observed in several other boreholes as far south as the Pencaitland area.

The interval between the Bilston Burn Limestone and the horizon of the Top Hosie Limestone consists almost entirely of medium- and fine-grained sandstone up to 30 m thick in the Bankhead area. The strata become more argillaceous eastwards with thinner sandstone bands and vary greatly in thickness from over 43 m in the north to less than 9 m in the south.

Top Hosie Limestone

With the exception of one small exposure of compact yellow limestone formerly seen near H.W.M. on the shore between Longniddry and Port Seton, tentatively correlated with the Top Hosie Limestone, information about this bed is entirely derived from boreholes. The thickest development recorded in the district was in a borehole put down underground in Fleets Colliery, south-east of Tranent, [4111 7190], in which 0.7 m of hard, grey, crinoidal limestone was encountered. The evidence indicates that the bed thins northwards, eastwards and southwards, and that in places the limestone is absent and the horizon is represented only by a thin marine band. At a few localities no evidence of marine conditions was recorded.

The variability of the sediments reflects the general change in conditions of depositon which took place at the end of Lower Limestone Group times. Shallow water prevailed, and the environment became less marine and more deltaic. WT, ADM

CHAPTER 6

Upper Carboniferous

INTRODUCTION

Sediments of Upper Carboniferous (Silesian) age are confined to the north-west part of the district. They are divided, in ascending order, into the Limestone Coal Group, the Upper Limestone Group, the Passage Group and the Coal Measures. The first two groups and part of the Passage Group are of Namurian age and the upper part of the Passage Group together with the Coal Measures belong to the Westphalian. Only the basal part of the Lower Coal Measures is present on land in the district. The zonation of the succession and the palaeontology are discussed in Chapter 7.

LIMESTONE COAL GROUP

The base of the group is drawn at the top of the Top Hosie Limestone and includes all the beds up to the base of the Index Limestone. The group is characterised by the presence of numerous coals and the succession in the district is shown on Figure 14 together with the correlation with the sequence in the Midlothian Coalfield to the west, where the group has a thicker development.

The outcrop of the Limestone Coal Group (Figure 15) crosses the coast at Seton Sands between Port Seton and Longniddry. Inland the group underlies a broad area from Prestonpans, Tranent, Macmerry and Gladsmuir in the north, through Elphinstone, Ormiston and Wester Pencaitland, to Pathhead in the south. From Seton Sands the outcrop of the base of the group trends generally south-east for 4 km to Gladsmuir, then south for 2 km to Nisbet Loanhead, where the strike of the basal beds assumes a south-westerly trend to the west margin of the district near Pathhead. The base of the group also crops out in the core of the D'Arcy–Cousland Anticline at Bankhead west of Tranent. The Index Limestone, at the top of the group, crops out in the deeper part of basin-structures lying between Seton and Elphinstone, and also round the plunging nose of the D'Arcy–Cousland Anticline between Prestonpans and Cockenzie.

The Limestone Coal Group strata of the East Lothian Coalfield have been gently folded into a complex syncline with dome and basin structures. The deepest basin structures occur to the north-west of the coalfield between Seton and Elphinstone, where some overlying Upper Limestone Group strata are preserved. Dips in the coalfield are in general very low, 5° or less. Steep dips occur close to faults or in folds of glaciotectonic origin (p. 76). To the west the East Lothian Coalfield is separated from the Midlothian Coalfield by the northward-plunging D'Arcy–Cousland Anticline, the axis of which is cut obliquely by the Crossgatehall Fault.

Within the district the full thickness of the Limestone Coal Group succession is preserved only in the centre of the deep basins in the north-west, and round the northerly-dipping nose of the D'Arcy–Cousland Anticline at Prestonpans. As the strata are covered almost entirely by Quaternary deposits, there are very few natural sections in the Limestone Coal Group. Poor coastal sections at the west end of Seton Sands expose some of the higher beds. Inland, only short sections of strata are present. Large temporary exposures at the Blindwells Opencast Site, near Tranent, show sections in the upper part of the succession from the Upper Craw Coal to the Great Seam Coal (Plates 7, 8).

Most information on the Limestone Coal Group has come from borehole and mining records. The only borehole in the district to pass through the complete succession was at Port Seton which proved a total thickness of 166 m for the group. Elsewhere records cover only part of the sequence, suggesting, nevertheless, that the group does not vary greatly in thickness throughout the district.

During Limestone Coal Group times, the district was part of a flat coastal area. In the predominantly deltaic environment a cyclic sequence of sandstones, siltstones and mudstones with coals and seatclays was deposited. This rhythmic sequence differs from that found in the Lower Limestone Group by the presence of more numerous and thicker coal seams and a major reduction in the development of marine sediments. Conditions were particularly favourable for establishment of vegetation on the swampy land surface, and for the accumulation and preservation of the vegetable matter, resulting in the formation of thick coal seams. Slight subsidence and the onset of near-shore marine conditions gave rise to mudstone horizons containing the brachiopod *Lingula* immediately overlying some of the coals. More prolonged subsidence gave rise to two marine bands containing a rich fauna of brachiopods and molluscs. These are the Johnstone Shell Bed and the Black Metals Marine Band which are present over most of central Scotland and are of primary importance in the correlation of the group.

DETAILS

Top Hosie Limestone to Lower Diamond Coal

The interval between the Top Hosie Limestone, at the base of the Limestone Coal Group, and the Lower Diamond Coal varies from 34 m near Tranent to 27 m in the eastern part of the coalfield. The strata include three thin, impersistent coal seams, the Hauchielin, Little and Andrews coals, and the Johnstone Shell Bed, the lower of the two marine bands in the group.

Hauchielin Coal

The Hauchielin Coal lies only 3 or 4 m above the Top Hosie Limestone. This seam is correlated with the Arniston Parrot Coal of the Newtongrange area on the eastern side of the Midlothian Coalfield. It is a generally thin seam, records of thickness ranging

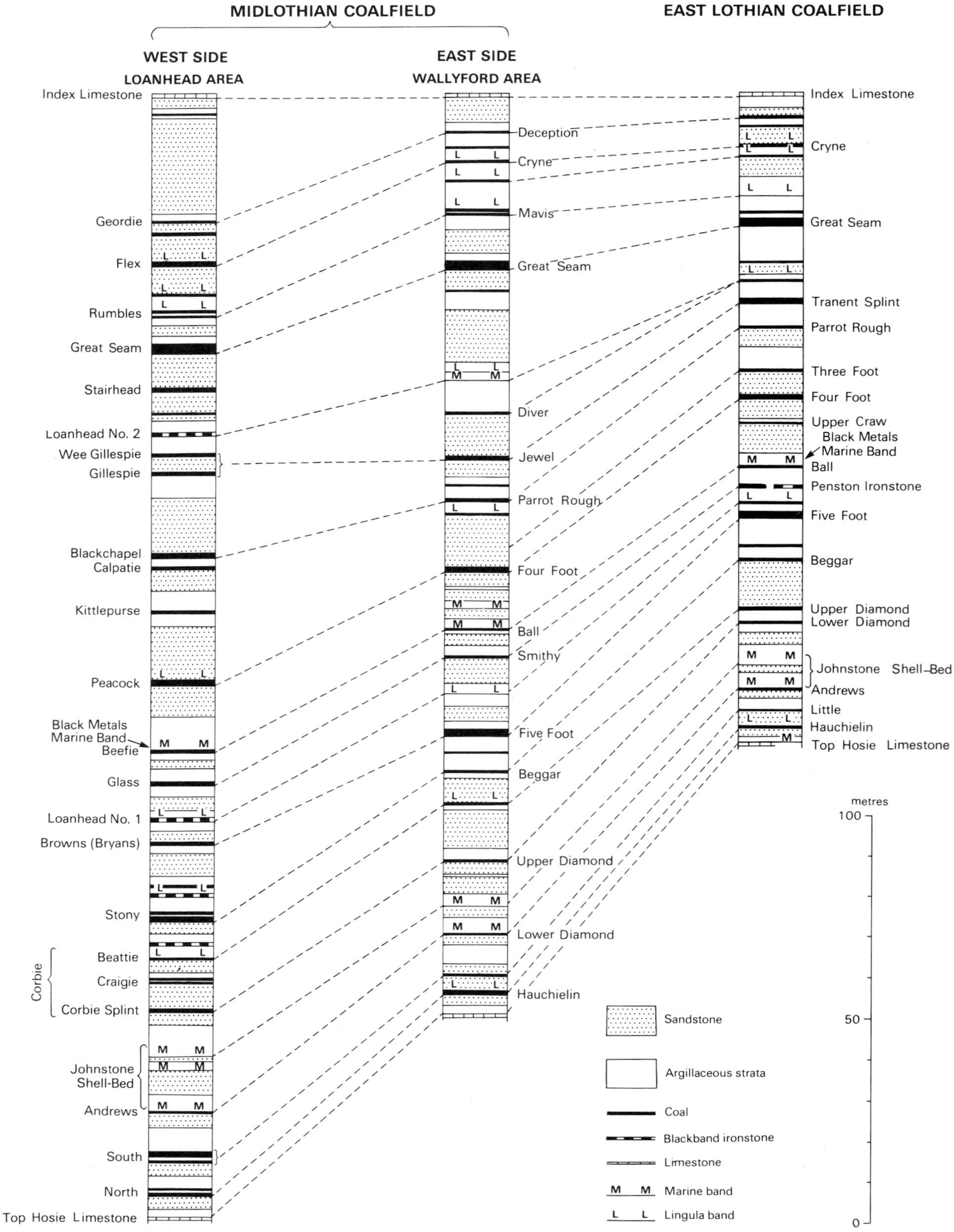

MIDLOTHIAN COALFIELD

WEST SIDE
LOANHEAD AREA

Index Limestone
Geordie
Flex
Rumbles
Great Seam
Stairhead
Loanhead No. 2
Wee Gillespie
Gillespie
Blackchapel
Calpatie
Kittlepurse
Peacock
Black Metals
Marine Band
Beefie
Glass
Loanhead No. 1
Browns (Bryans)
Stony
Corbie
Beattie
Craigie
Corbie Splint
Johnstone
Shell-Bed
Andrews
South
North
Top Hosie Limestone

EAST SIDE
WALLYFORD AREA

Deception
Cryne
Mavis
Great Seam
Diver
Jewel
Parrot Rough
Four Foot
Ball
Smithy
Five Foot
Beggar
Upper Diamond
Lower Diamond
Hauchielin

EAST LOTHIAN COALFIELD

Index Limestone
Cryne
Great Seam
Tranent Splint
Parrot Rough
Three Foot
Four Foot
Upper Craw
Black Metals
Marine Band
Ball
Penston Ironstone
Five Foot
Beggar
Upper Diamond
Lower Diamond
Johnstone Shell-Bed
Andrews
Little
Hauchielin
Top Hosie Limestone

metres
100
50
0

Sandstone
Argillaceous strata
Coal
Blackband ironstone
Limestone
M M Marine band
L L Lingula band

Figure 14 Correlation of the Limestone Coal Group in Midlothian and East Lothian

from 0.08 to 0.57 cm. It contains splinty bands but no gas coal for which the Arniston Parrot Coal was extensively exploited. The seam is not known to have been worked in the district. A thin bed of dark grey mudstone with *Lingula* and fish remains, directly overlying the Hauchielin Coal, has been recorded throughout the coalfield.

Little Coal

The Little Coal occurs about 3 to 5 m above the Hauchielin Coal. It is correlated with the South Coal of Midlothian. The seam is about 0.2 m thick in the south of the coalfield and absent in the north.

Andrews Coal

The interval between the Little and Andrews coals is generally about 5 m, but is 15 m at Fleets Colliery, near Tranent. The Andrews Coal position is defined by the overlying Johnstone Shell Bed. As much as 0.46 m thick at Bankton, near Tranent and south of Penston, the seam is more commonly only a few centimetres thick or absent.

Johnstone Shell Bed

The Johnstone Shell Bed in the district generally occurs as two beds of mudstone with marine fossils. The interval between the two

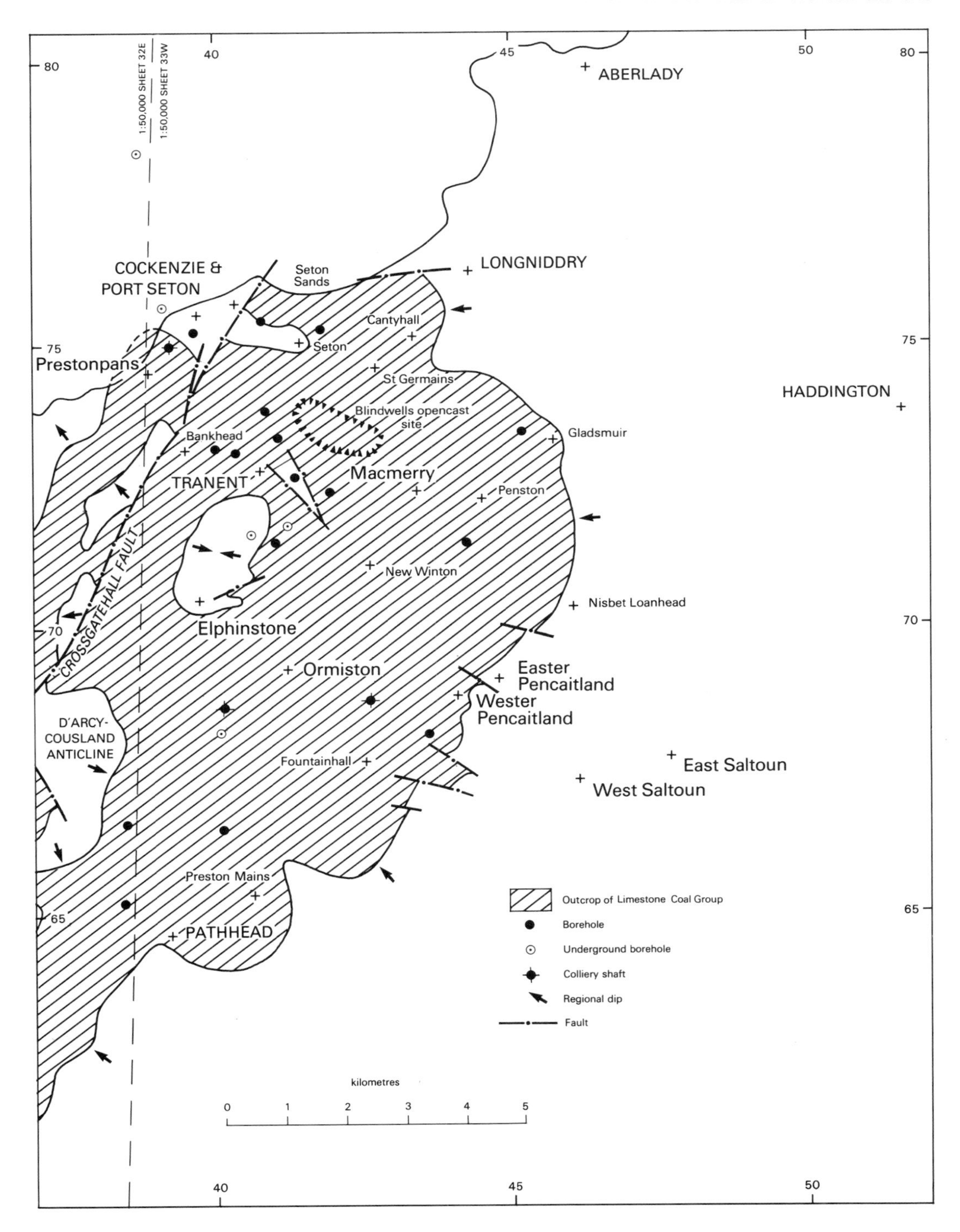

Figure 15 Generalised map of the Limestone Coal Group outcrops

Plate 7 Limestone Coal Group strata; till overburden on Great Seam down to Four Foot Coal, Blindwells Opencast Site, Tranent (D3482)

leaves ranges from 4 to 9 m. Both leaves contain bands and nodules of ironstone, and the lower part of each is the more fossiliferous. A thin band of impure marine limestone was recorded in the basal part of the lower leaf in a borehole at Hoprig, 1 km SE of St Germains. A similar thin marine limestone was found in the lower part of the upper leaf at a few localities. The two beds of mudstone are separated by a sandstone, the upper part of which in places is bluish, bioturbated and contains marine shells. At some localities a thin seam of coal underlies the upper leaf. A sandstone, 4 to 8 m thick, generally lies between the upper leaf and the Lower Diamond Coal.

Lower Diamond Coal and Upper Diamond Coal

These coals have received a variety of names in different parts of the district but the names Upper Diamond and Lower Diamond which were in use at Fleets Colliery are used here.

The Lower and Upper Diamond coals are correlated with the Corbie Group of three coals on the western side of the Midlothian Coalfield (Figure 14). The Lower Diamond Coal is the equivalent of the lowest seam, the Corbie Splint Coal, the Upper Diamond Coal correlates with the highest seam, the Beattie Coal, but there is no representative in the district of the middle seam, the Craigie Coal. The *Lingula* band which is a widely occurring marker band in the roof of the Beattie Coal in Midlothian has not been found in the present district. Both Lower and Upper Diamond coals were of fairly good quality and were used for household purposes and for steam-raising.

Lower Diamond Coal

The thickness of this seam varies considerably throughout the coalfield, ranging from 0.08 to 0.86 m. It was extensively mined where it was thick in the southern part of the coalfield, south and east of Ormiston, and in the north-east part between Seton Sands and the quartz-dolerite dyke at Cantyhall. In the middle of the coalfield it was only worked to a limited extent near Macmerry and from Fleets Colliery near Tranent.

Predominantly argillaceous strata lie between the Lower and Upper Diamond coals. The interval ranges in thickness from 3 to 6 m in the district.

Plate 8 Old stoop and room workings in the Great Seam, exposed in Blindwells Opencast Site, Tranent (D3484)

Upper Diamond Coal

As with the lower seam there is a considerable range of thickness for the Upper Diamond Coal, from 0.31 to 1.14 m. There were extensive workings of the seam in the north-eastern, eastern and southern parts of the coalfield, but in the deeper parts of the basin only limited workings were carried out.

Upper Diamond Coal to Five Foot Coal

The strata between the Upper Diamond and Five Foot coals are thickest, 30 m mainly of sandstone, in the western part of the coalfield, and thin to about 15 m of more argillaceous beds on the eastern side. The strata include three generally thin coals, the lowest of which is the Beggar Coal, and in places in the upper part an ironstone, the No. 2 Ironstone.

Beggar Coal

The Beggar Coal correlates with the Stony Coal of the Midlothian Coalfield. The seam is about 1 m thick at Prestonlinks Colliery [394 754], where it was worked under the Firth of Forth, and south of Prestonpans, where it was worked from the Dolphingstone Colliery [383 733] just beyond the western edge of the district. A coal 0.64 m thick, known as the Craw Coal or Lower Craw Coal, and mined to a very limited extent from two collieries at Macmerry, is possibly the equivalent of the Beggar Coal.

At a similar horizon is a coal seam, 0.8 m thick in two leaves, worked in a small area from Tynemount Colliery, south-west of Ormiston. At Oxenfoord, near Pathhead, the coal seam, 0.38 m thick, at this horizon is known as the Alecks Coal.

No. 2 Ironstone

At Dolphingstone Colliery the No. 2 Ironstone is a blackband ironstone, 0.69 m thick, lying 5.5 m below the Five Foot Coal. This ironstone has been correlated with a coal seam overlain by mudstone with beds of clay-ironstone proved in a borehole nearby. It has also been correlated with a musselband, composed mainly of carbonaceous shale with thin beds of clay-ironstone, an impure blackband ironstone, at Muirpark Farm, between Tranent and Macmerry.

Five Foot Coal

In economic importance the Five Foot Coal ranked next to the Great Seam and the Tranent Splint coals in the East Lothian Coalfield. Where best developed the Five Foot Coal is 1.5 m thick, but is more commonly 1 to 1.2 m thick, often in two or more leaves. In some places the coal thins to 0.2 m or is even completely replaced by sandstone.

North-west of the Crossgatehall Fault the Five Foot Coal was extensively worked, both inland and undersea from Prestonlinks Colliery, where it was 1.2 m thick. South-east of the fault it has been mined under most of the western part of the coalfield, north-west of Tranent from Bankton Colliery where is was also 1.2 m thick, and in the Tranent – Elphinstone – Ormiston area from several collieries where the seam was only 1 m thick. In the south-west the coal is over 1 m thick at Oxenfoord, but thins rapidly in places. In the north-east the Five Foot Coal is over 1 m thick in two leaves at Seton Mains, but other boreholes indicate the seam thins eastwards. In the Penston and Fountainhall areas on the east of the coalfield, the Five Foot Coal was much more variable in thickness and quality than further west. At Fountainhall the seam was 1.5 m thick, but described as foul coal. Coal was taken to Prestonpans for use as fuel to evaporate seawater in saltpans for the production of salt. More generally the seam was used for steam-raising, gas manufacture and as household coal.

Five Foot Coal to Ball Coal

The interval between the Five Foot Coal and the Ball Coal is generally about 10 m. The strata thicken to 14 m in the north at Prestonlinks and Seton and in the east at Macmerry, and locally they are 22 m thick in the south. This interval includes two thin coal seams, both of which locally develop ironstones. The lower has a *Lingula*-band in the roof; the upper is known as the Smithy Coal or the Penston Ironstone. The *Lingula*-band lies a few metres above the Five Foot Coal and is correlated with the widespread *Lingula*-band in the northern part of the Midlothian Coalfield on the horizon of the Loanhead No. 1 Ironstone (Tulloch and Walton, 1958, p. 42).

Smithy Coal and Penston Ironstone

A thin coal up to 0.48 m thick in leaves is present, several metres above the *Lingula*-band, in the western edge of the district and is correlated with the Glass or Smithy Coal of the Midlothian Coalfield (Figure 14). Locally developed on this horizon in the Penston – Macmerry area is a blackband ironstone, about 0.35 m thick which was mined locally. One record at Penston gave a thickness of 0.39 m of ironstone in two leaves. Recent shallow boreholes just south of the A1 road near Penston and Macmerry encountered workings in the Penston Ironstone and have allowed comparison with the Dolphingstone succession. In one borehole the workings were encountered 3 m above the *Lingula*-band which correlates with the No. 1 Ironstone at Dolphingstone. Thus the Penston Ironstone is not the equivalent of the Dolphingstone No. 1 Ironstone as previously suggested (Anderson *in* Clough and others, 1910, p. 194; Macgregor and others, 1920, pp. 179 – 181).

Ball Coal to Tranent Splint Coal

The interval between the Ball and Tranent Splint coals is generally about 40 m but ranges in thickness from a maximum of over 50 m in the Oxenfoord area in the south-west to a minimum of 30 m at Seton in the north. The strata include at the base the important Black Metals Marine Band, underlain by the Ball Coal; a thin local seam, the Upper Craw Coal; and four thick economic coals, the Four Foot, Three Foot, Parrot Rough and Tranent Splint coals.

Ball Coal

The Ball Coal is defined by the overlying Black Metals Marine Band. The seam is rarely more than 0.6 m thick, in places occurring in several leaves. The only records of workings are from boreholes east of Macmerry suggesting that the Ball Coal was mined along with the underlying Penston Ironstone.

Black Metals Marine Band

Throughout the East Lothian Coalfield the Ball Coal is overlain by a bed 3 m thick, of dark grey mudstone and silty mudstone with nodules and bands of clay-ironstone. This bed contains marine fossils, in all sections which have been examined in detail, which enable correlation with the Black Metals Marine Band, the upper of two major marine bands of the Limestone Coal Group in central Scotland (Wilson, 1967, p. 449). The fauna consists mainly of brachiopods and bivalves and is similar to that found in the Johnstone Shell Bed, in the lower part of the group, although the Black Metals Marine Band generally contains a less varied assemblage.

The beds between the marine band and the Four Foot Coal normally include one or more thick sandstones. Around Tranent a coal, up to 0.25 m thick, the Upper Craw Coal, is developed 4 or 5 m below the Four Foot Coal. This seam is the lowest worked coal in the Blindwells Opencast Site (Plate 7). Overall the interval between the Black Metals Marine Band and the Four Foot Coal ranges in thickness from 13 to 22 m in the district.

Four Foot Coal

This seam is the same as the Four Foot Coal on the east side of the Midlothian Coalfield and the Peacock or Blackchapel Coal on the west side (Figure 14). The seam is thickest in the west of the coalfield, 1.6 m in three leaves at Prestonlinks and 1.6 m in three leaves at Oxenfoord, thinning to 0.5 m at Tranent. The Four Foot Coal was worked only to a limited extent in the northern part of the coalfield, including undersea workings from Prestonlinks Colliery. There the seam was also known as the Third Fireclay Seam, from a 0.6-m thick fireclay parting which was mined. The Four Foot Coal was more extensively wrought in the central and southern parts of of the coalfield. The coal was used for steam-raising and for household purposes.

The Four Foot Coal is separated from the overlying Three Foot Coal by 3.5 to 5 m of mainly sandy strata.

Three Foot Coal

The Three Foot Coal is absent in parts of the eastern Midlothian Coalfield, but it is correlated with the Kittlepurse Coal (Figure 14), which is developed elsewhere in the coalfield. At its thickest the Three Foot Coal is just over 1 m in the north of the district. Normally a single seam, it splits into leaves in the south. The Three Foot Coal has been fairly extensively worked in the northern and central parts of the coalfield. In the south the seam was worked in a small syncline near Oxenfoord, where it was 0.86 m thick in two leaves. Use of the Three Foot Coal at Prestonlinks Colliery was for steam-raising and gas manufacture.

The interval between the Three Foot Coal and the Parrot Rough Coal is 15.5 m at Bankton where a thick sandstone with a coarse, gritty, basal part is developed but generally it is 5 to 10 m and only 2 m at Oxenfoord in the south.

Parrot Rough Coal

The Parrot Rough Coal, also known as the Parrot Coal, is the equivalent of the Blackchapel Coal in the Midlothian Coalfield (Figure 14). As the name indicates, the seam in places contains a band of gas coal 0.1 to 0.2 m thick, in several parts of the district. In thickness the seam is generally 0.5 to 0.9 m, but is 1.22 m thick offshore at Prestonlinks, and is also over 1 m thick in a small syncline at Oxenfoord in the south. In many places the seam occurs in leaves and there were fairly extensive workings in the Oxenfoord area.

Normally the Parrot Rough Coal and the Tranent Splint Coal are separated by only a few metres of mainly argillaceous strata.

Tranent Splint Coal

Correlated with the Gillespie Group of coals and the Jewel Coal of the Midllothian Coalfield (Figure 14), the Tranent Splint Coal ranked next to the Great Seam in economic importance in the district. Records of the seam's thickness varied from 0.33 m to as much as 1.8 m and the coal occurred in two or more leaves. The seam was worked extensively in many parts of the district and was used for household coal and for steam-raising.

Tranent Splint Coal to Great Seam

This section of strata between the two most important coal seams in the district ranges from 30 m at Prestonlinks Colliery where the strata are mainly sandstone and include the Diver Coal, to 16.5 m in the Port Seton to Seton area. Near Tranent the seams are separated by 17.5 m of strata and at Elphinstone by 20 m of strata. One or two thin coal seams are present locally, the lowest with a *Lingula*-band in the roof.

The *Lingula*-band lies from 0.53 m to over 6 m above the Tranent Splint Coal, in the Tranent area. This *Lingula*-band corresponds to the one found in Midlothian in the Gillespie Group and on the horizon of the Loanhead No. 2 Ironstone and to the band with *Lingula* and *Orbiculoidea* at Wallyford (Figure 14).

Diver Coal

This good quality coal, 0.66 to 0.84 m thick, occurs 12 m above the Tranent Splint Coal within a thick sandstone sequence at Prestonlinks and elsewhere in the north of the district.

Great Seam

The Great Seam is the thickest coal in the Limestone Coal Group of the district, and is recognised under this name, virtually throughout the East Lothian and Midlothian coalfields. The seam has been worked extensively both inland and undersea. It is commonly around 2 m in thickness, usually with a thick lower leaf and one or two thinner upper leaves.

Great Seam to Index Limestone

The interval between the Great Seam and the Index Limestone is greater to the north-west of the Crossgatehall Fault, being about 40 m in boreholes just west of the district boundary. East of the dislocation the strata thin to 29 m at one locality. Elsewhere there are no complete sections through this interval but strata belonging to it crop out on the shore west of Seton Sands. The detailed succession between the Great Seam and the Index Limestone also varies considerably. Several coal seams occur, including representatives of the Mavis, Cryne, and Deception coals of the Wallyford area in Midlothian, correlation being based on the presence of the three *Lingula*-bands. In the present district these coals are generally thin, with the exception locally of the Cryne Coal.

Cryne Coal

Correlated with the Cryne Coal at Wallyford, and the Flex Coal in the west of the Midlothian Coalfield, this seam was locally thick enough to have been worked. Two boreholes in the Tranent area recorded wastes at this horizon.

UPPER LIMESTONE GROUP

The base of the group is taken at the base of the Index Limestone, which is well developed in the district. In central Scotland generally there are four principal limestone horizons, the Index, the Orchard, the Calmy and the Castlecary limestones, the top of the last being taken as the top of the group. The Orchard Limestone is termed the Orchard Beds in Midlothian and East Lothian. In the present district, as in parts of the Midlothian Coalfield (Tulloch and Walton, 1958, pp. 61, 80), the Castlecary Limestone is thought to be absent due to an unconformity cutting out the upper strata of the Upper Limestone Group. In addition, there is a general eastward thinning of the group from Midlothian to East Lothian.

The Upper Limestone Group outcrop lies along the eastern limb of the Midlothian basin, round the nose of the faulted D'Arcy–Cousland Anticline and in the Port Seton Syncline. Exposures of Upper Limestone Group strata are confined to the foreshore east of Port Seton Harbour (Figure 16), where the beds dip to the SW or WSW on the east limb of the syncline, at angles ranging from 18° to 40°. The strata are cut by several faults and just south of the shore, they are truncated by the Morton Vale Fault. On the west side of the syncline, Upper Limestone Group strata underlying the Preston Links area south-west of Cockenzie were formerly exposed south-west of Cockenzie Harbour, but are now obscured by the Cockenzie Power Station.

Several small inliers of the Upper Limestone Group, concealed by superficial deposits, have been proved by boreholes in the north-west part of the East Lothian Coalfield. The largest extends from Tranent 2.5 km SW to Elphinstone, while five smaller faulted inliers lie between Port Seton and Tranent. Only the basal part of the group occurs in them.

The Upper Limestone Group strata were deposited in an alternating marine and deltaic environment, similar to that of the Lower Limestone Group. The group is composed of cyclic sequences of sandstones, siltstones, mudstones and thin marine limestones, with seatclays and several thin coals. The section at Port Seton shore is dominantly arenaceous, with several thick sandstone beds.

DETAILS

Index Limestone

The Index Limestone is well known from boreholes, but in the district is exposed only at Port Seton shore (Figure 16). It reaches its maximum known thickness, 1.12 m, at Preston Links, south-west of Cockenzie. At Port Seton shore, 2 km to the east, the limestone is only 0.23 to 0.28 m thick. Inland the limestone ranges from 0.31 to 0.46 m in thickness in the inlier south-east of Port Seton, while in the large inlier between Tranent and Elphinstone the limestone reaches 1.07 m. The limestone is grey, rather argillaceous and sandy in parts, hard when fresh, but found to be softer and decalcified in some boreholes. Algal patches, shell fragments and crinoid debris are present, but the fauna is poor.

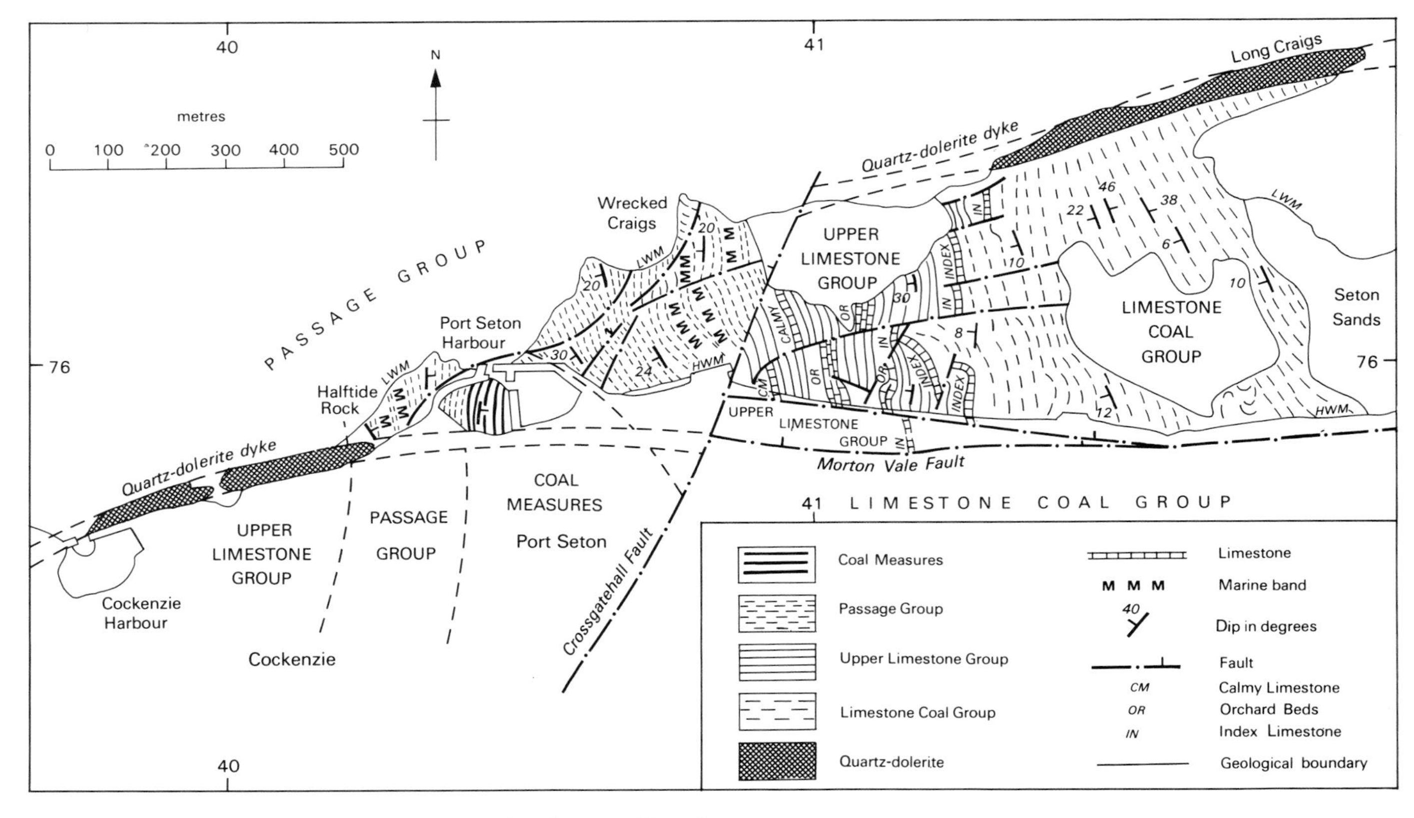

Figure 16 Map of the solid geology on the shore at Port Seton

Index Limestone to South Parrot Coal

This interval is consistently 25 to 30 m thick on the east side of the Midlothian Coalfield and in the district. Beds of mudstone with marine shells lie immediately above the Index Limestone, but most of the strata consist of grey-white, fine- to medium-grained sandstone.

South Parrot Coal

Reaching 1 m or more in parts of the Midlothian Coalfield, the South Parrot Coal consists of 0.3 m of bright and dull banded coal at Preston Links. Several coal positions occur in the strata immediately above. No coals have been recorded at Port Seton shore, probably because they are concealed by beach deposits rather than absent.

South Parrot Coal to Orchard Beds

This interval is 70 m thick in the eastern part of the Midlothian Coalfield but only about 30 m of poorly exposed strata with gaps are seen on Port Seton shore where only white, grey and yellow, cross-bedded sandstone beds are seen.

Orchard Beds

The section of the Orchard Beds in an offshore borehole near the western margin of the district closely resembles sections of these beds in the Midlothian Coalfield. A lower grey, shelly and crinoidal earthy limestone, 0.38 m thick, is separated from a higher, grey, shelly and crinoidal, fine-grained limestone, 0.5 m thick, by 5.73 m of grey calcareous mudstone and shale with thin limestones in the upper part. The upper bed of limestone is overlain by 6.96 m of grey mudstone and dark grey shaly mudstone with crinoid debris and shell fragments in the lower part and ironstone bands and nodules in the upper part. On the shore east of Port Seton Harbour the Orchard Beds are only partly exposed, but two beds of limestone are seen. The lower limestone is grey, compact and lenticular, ranging up to 0.28 m thick, and contains a rich marine fauna of corals, brachiopods including *Semiplanus* cf. *latissimus* and molluscs, typical of this horizon in Midlothian (Tulloch and Walton, 1958, pp. 66–72). The upper limestone is hard, grey, fine-grained, with crinoid debris, and is about 0.3 m thick. The intervening strata, estimated to be 7.3 m thick, are partly obscured, but include fine- to medium-grained, grey-white sandstone and grey siltstone. The upper of these two thin limestones was formerly identified as the Calmy (Arden) Limestone (Bailey *in* Clough and others, 1910, p. 155), but palaeontological evidence indicates that it is the upper leaf of the Orchard Beds.

Orchard Beds to Calmy Limestone

On the shore east of Port Seton Harbour, the interval between the Orchard Beds and the Calmy Limestone includes two thick sandstones, but there is a large gap with no exposures in the basal part of the sequence. The strata at Port Seton have been estimated to be 39 m thick.

Calmy Limestone

A metre thick just beyond the western edge of the district, the Calmy Limestone is a compact fine-grained bed with crinoid debris, only 0.38 m thick at Port Seton shore, the only place where the limestone is seen at outcrop in the district. This limestone was formerly identified as the Castlecary Limestone, but the underlying silty mudstone has yielded a marine fauna, including *Actinopteria regularis*, *?Edmondia punctatella* and *Sanguinolites* cf. *clavatus*, all forms characteristic of the Calmy Limestone in Midlothian (Wilson *in* Tulloch and Walton, 1958, p. 81).

Strata above the Calmy Limestone

On the shore east of Port Seton Harbour only about 35 m of poorly exposed Upper Limestone Group strata are present above the Calmy Limestone. Formerly it was believed that the Upper Limestone Group passed conformably up into the Passage Group (Clough and others, 1910, p. 158), but it is now recognised that east of Port Seton Harbour the Upper Limestone Group and the Passage Group are separated by a fault. In the Cockenzie and Port Seton area it is thought that the Castlecary Limestone and the uppermost strata of the Upper Limestone Group are cut out by an unconformity. Elsewhere along the eastern limb of Midlothian Basin and in most of the Penicuik Syncline, the Castlecary Limestone is also absent (Tulloch and Walton, 1958, pp. 61, 80 and pl. 4).

PASSAGE GROUP

The term Passage Group was proposed by MacGregor (1960) to include the strata from the top of the Castlecary Limestone, the top of the Upper Limestone Group, up to the base of the Coal Measures in Scotland. The base of the Coal Measures (Westphalian) in western Europe is drawn at the Subcrenatum Marine Band but this band has not been found in Scotland. Palaeontological evidence indicates that the horizons taken as the local base of the Coal Measures on lithological grounds in various places in central Scotland are younger than the Subcrenatum Marine Band. The Passage Group therefore includes the Namurian – Westphalian boundary.

In the present district the Castlecary Limestone is missing as a result of local unconformity and the base of the Passage Group is taken at the top of the highest beds of the Upper Limestone Group. The top of the group is drawn at the base of the strata interpreted as Lower Coal Measures.

Passage Group strata are present only in a small area be tween Port Seton and Bankton. At Port Seton the beds are preserved in the core of the Port Seton Syncline which plunges south-south-east and is truncated on the south-east side by the Crossgatehall Fault (Figure 16). Exposures are confined to the foreshore at Port Seton. The strata are well exposed on the eastern limb of the syncline, north-east of Port Seton Harbour, where the beds dip to the SW at angles of 10 to 30°. A less extensive section is seen west of the harbour on the western limb of the fold on which the strata dip eastwards at 20° to 25°. Inland, borehole evidence indicates that the fault-bounded outcrop extends under the superficial cover to Bankton.

The succession at Port Seton, estimated at about 100 m thick, is composed predominantly of sandstones, red-stained in parts with ochreous weathering, coarse and pebbly in places with conglomeratic bands. Beds of grey mudstone and silty mudstone, locally stained purple and red, are also present, along with grey, purple and red, mottled seatclays and a few thin coal seams. Two marine bands are present in the lower part of the succession, the lower being a nodular, red, calcareous ironstone about 0.3 m thick, with marine fossils in a shaly matrix, and the upper a grey, purple-and red-stained mudstone, 1.5 m thick, with ironstone nodules and *Productus*. These two marine bands, separated by 11 m of mainly massive white-grey sandstone, are exposed both east and west of Port Seton Harbour. Neither marine band can be correlated with any certainty outside the district. Nodules and bands of red- and purple-stained ironstone occur at several horizons.

The predominance of beds of sandstone, coarse and pebbly in parts, suggests that the Passage Group was deposited in a dominantly fluviatile environment, while the two marine bands are evidence of periodic incursions of the sea. The red, purple and yellow staining which affects part of the succession is considered to be of secondary origin, but it is not clear whether this coloration dates from intra- or post-Carboniferous times.

LOWER COAL MEASURES

MacGregor (1960) subdivided the Scottish Coal Measures into Lower, Middle and Upper divisions but in the district only part of the Lower Coal Measures is preserved. The Subcrenatum Marine Band, the base of which is taken as the base of the Westphalian in England and Wales (Ramsbottom and others, 1978, p. 45), has not been recognised in Scotland. This marine band lies at the base of the Lenisulcata Chronozone, the basal part of the Lower Coal Measures. The local base of the Coal Measures in Scotland is taken at the lowest convenient mappable horizon in each coalfield, which in Midlothian and East Lothian is the base of the Seven Foot Coal or the equivalent Melville Group of coals (Tulloch and Walton, 1958, p. 93). As far as is known the basal beds of the Lower Coal Measures in these areas belong to the higher part of the Communis Chronozone, hence the true base of the Westphalian probably lies in the upper part of the Passage Group.

In the district, Coal Measures strata are confined to a small area in the vicinity of Port Seton where the basal 30 m or so of the Lower Coal Measure are preserved in the core of the small, SSE-plunging Port Seton Syncline. The rocks are exposed only within Port Seton Harbour, where the general dip is 30° eastwards on the west limb of the syncline. The upper part of the succession is even less well exposed than formerly. Inland the outcrop, concealed by superficial deposits, extends for 1 km, being truncated to the south-east by the Crossgatehall Fault and to the south by a WSW-trending fault.

During Coal Measures times, conditions of deposition were similar in many respects to those prevailing during the deposition of the Limestone Coal Group. The environment was dominantly deltaic, and the presence of thick coal seams indicates prolonged accumulation of vegetation under swamp conditions. Marine incursions occurred very rarely in the recurring cycles of sedimentation. Non-marine bivalves (mussels) lived in the shallow waters and their shells are preserved as musselbands above some coal seams.

DETAILS

This lowest part of the Lower Coal Measures is the only part of the Coal Measures preserved in the district. Knowledge of the sequence is based almost entirely on the section in Port Seton Harbour, recorded by Bailey in 1903 (*in* Clough and others, 1910, p. 160), and now poorly exposed. The Port Seton section was:

	Thickness m
Sandstone, cross-bedded, coarse-grained	1.8 +
NINE FOOT COAL	1.93
Seatclay, grey, rooty, ironstone nodules	2.75
Mudstone, dark grey, prominent ironstone nodules	0.9
Ironstone, nodular	0.15
PINKIE THREE FOOT COAL	0.1
Seatclay, dark grey	0.3
Mudstone, dark grey, sandy ironstone bands	1.5
Seatclay, grey	0.15
SIX FOOT COAL	2.13
Seatclay, rooty	0.75
Siltstone, grey, micaceous, beds of rooty seatclay, irony nodules and bands	6.25
Gap with foul coal	1.5
Sandstone, flaggy, and silty mudstone	5.15
Mudstone, sandy, dark grey, well bedded	1.85
Mudstone, dark grey, ironstone nodules containing *Carbonicola* cf. *communis*	1.2
Gap with coal and parrot coal (positions of FOUR FOOT COAL AND SEVEN FOOT COAL)	1.5

In Midlothian, four thick coal seams occur in this part of the succession, the Seven Foot, Four Foot, Fifteen Foot and Nine Foot coals in ascending order. In the Port Seton section the lower two seams are not exposed, but the upper two are present, the Fifteen Foot Coal being split into two seams, the Pinkie Three Foot Coal (0.1 m) and the Six Foot Coal (2.13 m); and the Nine Foot Coal (1.93 m). The thickness of the sequence at Port Seton is about 30 m, with the strata between the coals being mainly argillaceous. The sequence thickens to over 40 m at Musselburgh to the west of the district.

Two musselbands, both belonging to the Communis Chronozone (Calver *in* Tulloch and Walton, 1958, p. 115), have been identified in this succession in parts of Midlothian. The upper musselband, above the Fifteen Foot Coal or the Pinkie Three Foot Coal, has long been recognised, while the lower, above the Four Foot Coal, has only more recently been discovered. The lower musselband is represented at Port Seton by the mudstone with *Carbonicola* cf. *communis* near the base of the succession, whereas the upper band is correlated with a mudstone with prominant ironstone nodules which overlies the thin Pinkie Three Foot Coal. No non-marine bivalves were recorded in this bed when the section at Port Seton was measured, and obscuring of the section prevents a further search. Bailey (*in* Clough and others, 1910, p. 160) correlated the musselband at the base of the Port Seton section with the one above the Fifteen Foot Coal in Midlothian, as at the time it was the only such band known in the succession. With recognition of two musselbands and the correlation revised, the Port Seton succession is seen to bear a close resemblance to the lower part of the Coal Measures in Midlothian. It seems probable that the Seven Foot and Four Foot coals are separated by only a small interval at Port Seton and that the gap in the section under the musselband conceals the outcrop of both seams. WT, ADM

CHAPTER 7

Carboniferous palaeontology

Fossiliferous strata of Carboniferous age in the district are now relatively poorly exposed as many of the limestone quarries which formerly yielded abundant specimens are now overgrown or filled in. Good exposures of some parts of the succession are still available on the shore in the Aberlady and Port Seton areas but in each case the succession is disturbed by faulting.

Resulting from the first examination of the ground by the Geological Survey, the first account of the Carboniferous fossils in the area was given by Salter (*in* Howell and others, 1866, pp. 70–77). A more comprehensive list of the species recorded was published in Clough and others (1910, pp. 206–217), based on collections made during the first revision of the area. In both of these works, information was restricted to the fossiliferous beds exposed at outcrop and there was no knowledge of the fossils present in some parts of the succession such as the Limestone Coal Group and parts of the Calciferous Sandstone Measures.

Since then, much of the sequence not exposed at outcrop has been drilled by boreholes and our knowledge of the fossils present in these beds is reasonably satisfactory. In particular, the IGS borehole at Spilmersford provided a sequence of the fossiliferous beds through all of the Calciferous Sandstone Measures and part of the Lower Limestone Group (Wilson *in* Davies, 1974, pp. 25–30). The fossils present in the Limestone Coal Group have also become known by the examination of borehole cores since the last revision of the district.

The details of the Lower Carboniferous marine faunas of the area were included in a paper describing these forms from the whole of south-east Scotland (Wilson, 1974). In this present general account only the principal elements of the faunas are discussed.

CLASSIFICATION AND ZONATION

The Lower Carboniferous (Dinantian) rocks of the district are divided into the Calciferous Sandstone Measures and the Lower Limestone Group. The Upper Carboniferous (Silesian) strata comprise the Limestone Coal Group, the Upper Limestone Group, the Passage Group and the Lower Coal Measures. Of the Lower Carboniferous divisions, the Calciferous Sandstone Measures form a very thick sequence of sediments and Wilson (1974, p. 38) proposed a subdivision into Lower and Upper Lothian groups with the boundary at the Macgregor Marine Bands. This subdivision is not used here, however, as it proved to be impractical for mapping purposes.

Biostratigraphical classification of the Scottish Carboniferous in the schemes used in England and Wales is difficult to apply as the diagnostic fossils are seldom present in the Scottish sediments. In the Lower Carboniferous the Lower Limestone Group was equated with the Upper Bollandian Stage (P_2) of the goniatite zonation (Currie, 1954, p. 534) and was placed in the Brigantian Stage of the classification of George and others (1976, p. 47). The Macgregor Marine Bands, which contain the lowest, rich marine fauna in the district, were placed in the Asbian Stage of the latter classification. Lack of diagnostic marine fossils below this level, however, precludes meaningful extension of this classification, to the lower, and major, part of the Calciferous Sandstone Measures.

One group of fossils, the miospores, is present throughout most of the Lower Carboniferous succession and a classification, using them, of the sequence proved in the Spilmersford Borehole was proposed by Neves and Ioannides (1974). Five miospore assemblage zones, based on the ranges of miospore species, were defined and these formed the basis of the miospore zonation of the Lower Carboniferous of Scotland and northern England of Neves and others (1973).

The Upper Carboniferous rocks cover only a very small part of the district and their zonation can only be inferred from other parts of Scotland. The Pendleian (E_1) Stage of the Namurian ranges from the top of the Lower Limestone Group up to the Orchard Beds (Currie, 1954; Ramsbottom, 1977) and thus includes the Limestone Coal Group and part of the Upper Limestone Group. The Arnsbergian (E_2) Stage includes strata from the Orchard Beds up to an horizon in the Passage Group which has not been defined in the area. The only Westphalian beds recorded are in a small area at Port Seton and are assigned to the Communis Chronozone of the Lower Coal Measures.

FOSSILS

Calciferous Sandstone Measures

The fossils present in the Calciferous Sandstone Measures indicate that there was a gradual change from non-marine to predominantly marine conditions over the district during the period of deposition. In the Spilmersford Borehole approximately 530 m of Calciferous Sandstone Measures were recorded but in the lowest 300 m practically no marine fossils were found. The fossils present in these beds were the plants *Alcicornopteris sp.*, *Cardiopteris sp.* and *Sphenopteris affinis* Lindley & Hutton, the worms *Serpula sp.* and *Spirorbis sp.*, the gastropod *Naticopsis? scotoburdigalensis* (Etheridge jun.), the bivalves *Lithophaga* cf. *lingualis* (Phillips), *Modiolus* cf. *latus* (Portlock) and *Naiadites* cf. *obesus* (Etheridge jun.), ostracods, arthropod fragments and fish remains including *Rhizodus sp.*

In the interval between this non-marine sequence and the base of the Macgregor Marine Bands, the fossils point to a non-marine facies except for those in two beds, the Dump and Limekiln marine bands which both contain poor marine faunas. Of the Calciferous Sandstone Measures succession occurring in the Spilmersford Borehole, slightly more than the lower two-thirds of it contains non-marine fossils except for two poorly developed marine bands near the top of this lower part. The sediments exposed on the shore in the North Berwick area also contain non-marine fossils and probably belong to the lower part of the Calciferous Sandstone Measures. The most famous fossiliferous locality on this coast is at Cheese Bay, about 3 km NE of Gullane, where a rich fauna of arthropods and fish was discovered by the Geological Survey at the beginning of the century (Peach *in* Clough and others, 1910, pp. 215–217).

The Macgregor Marine Bands are the lowest beds in the Carboniferous of the district which contain a rich marine fauna and the beds can be traced over much of south-east Scotland (Wilson, 1974). The type sequence is that found in the Spilmersford Borehole and comprises the Saltoun Marine Band, the Tyneholm Beds and the Winton Marine Band. The Sandersdean Limestone of the Haddington district is also of the same age. The combined fauna of the beds in the district includes the bryozoan *Stenodiscus haddingtonensis* (Lee), the brachiopods *Lingula spp., Pleuropugnoides sp., Productus sp., Pugilis sp.* and *Spiriferellina sp.*, the gastropod *Euphemites* cf. *urii* (Fleming), the bivalves *Aviculopecten subconoideus* Etheridge jun., *Edmondia sulcata* (Fleming), *Limipecten sp., Pteronites* cf. *angustatus* McCoy, *Schizodus pentlandicus* (Rhind) and *Streblopteria? redesdalensis* (Hind), goniatite fragments and the echinoid *Archaeocidaris sp.*

In the uppermost part of the Calciferous Sandstone Measures there is a group of marine bands comprising the Spilmersford Beds and the Lower and Middle Longcraig limestones. The Spilmersford Beds contain a relatively poor fauna of brachiopods and molluscs but the other two bands have yielded rich assemblages of brachiopods with corals, bryozoa and molluscs also present. The Lower Longcraig Limestone is distinguished by the presence of *Semiplanus* cf. *latissimus* (J. Sowerby) and the Middle Longcraig Limestone contains the colonial coral *Lithostrotion junceum* (Fleming) and large numbers of the brachiopods *Composita* and *Pleuropugnoides.*

Lower Limestone Group

The rich marine faunas present in the uppermost beds of the Calciferous Sandstone Measures continue into the lower part of the Lower Limestone Group and mark the period when marine influences were at their maximum over the district in Carboniferous times. The Upper Longcraig Limestone, at the base of the Lower Limestone Group, contains a very rich and varied fauna of corals, bryozoa, brachiopods and molluscs. The overlying Lower Skateraw Limestone does not have such a rich fauna but is distinguished by containing the maximum development of *Gigantoproductus* cf. *giganteus* (J. Sowerby) in the succession. The succeeding Skateraw Limestone contains the most varied fauna in the succession. A band of the foraminifer *Saccamminopsis fusulinaformis* (McCoy) occurs in the upper part of the limestone, the sponge *Hyalostelia parallela* (McCoy), the rare coral *Microcyathus* cf. *cyclostomus* (Phillips) and numerous species of brachiopods and molluscs are also present.

The succession in the upper part of the Lower Limestone Group is poorly known. Marine fossils have been recorded from thin limestones in a few boreholes but the information is too poor to correlate these beds with possible equivalents outside the district.

Limestone Coal Group

The only information concerning the fossils present in the Limestone Coal Group has been obtained from a few boreholes. Both of the major marine bands of the group, the Johnstone Shell Bed and the Black Metals Marine Band, are present and both contain similar faunas of brachiopods such as *Lingula squamiformis* (Phillips) and *Pleuropugnoides sp.* but the most abundant group is the bivalves such as *Aviculopinna* cf. *mutica* (McCoy) and *Streblopteria ornata* (Etheridge jun.). The Johnstone Shell Bed normally contains the more varied assemblage with bryozoa, *Productus sp.* and crinoid columnals present at some localities. *Lingula* bands are also present above some of the coals.

Upper Limestone Group

Three of the major marine horizons of the Upper Limestone Group are present on the shore east of Port Seton harbour. The lowest one, the Index Limestone, is not well developed and only a poor marine fauna has been obtained from it. The succeeding major marine sequence, the Orchard Beds, contain a rich marine fauna of corals, brachiopods including *Semiplanus* cf. *latissimus* and molluscs typical of the horizon in Midlothian (Tulloch and Walton, 1958, pp. 66–73). The uppermost band, the Calmy Limestone, is distinguished by the underlying silty mudstone which has yielded *Actinopteria regularis* (Etheridge jun.), *?Edmondia punctatella* (Jones) and *Sanguinolites cf. clavatus* (Etheridge jun.) which are all forms which characterise this horizon in Midlothian (Wilson *in* Tulloch and Walton, 1958, p. 81).

Passage Group

Two poorly fossiliferous marine bands occur above the Calmy Limestone at Port Seton. The upper one contains red-stained *Productus* cf. *carbonarius* de Koninck in a grey mudstone matrix and both bands probably belong to the Passage Group but cannot be correlated with any certainty outside the area.

Coal Measures

A bed formerly exposed inside Port Seton harbour was found to contain *Carbonicola* cf. *communis* Davies and Trueman. This is the only record of Westphalian strata in the district.

RBW

CHAPTER 8

Garleton Hills Volcanic Rocks

INTRODUCTION

The volcanic rocks within the Calciferous Sandstone Measures of the district reach their maximum development in the vicinity of the Garleton Hills and have become known as the Garleton Hills Volcanic Rocks. They are also present, but concealed, under younger sediments in western parts of the district. Though not continuing beyond the D'Arcy–Cousland anticline, apart from pyroclastic horizons, the rocks are approximately contemporaneous with the Arthur's Seat Volcanic Rocks of the Edinburgh district (Mitchell and Mykura, 1962, p. 45).

Upton (*in* Sutherland, 1982, p. 263) pointed out that the Garleton Hills lie over the subsurface extension of the main Southern Upland Fault-zone. He suggested that a magma chamber may have developed along this zone of weakness large enough for it to have produced basic magmas which gave rise, by fractionation, to trachytes and phonolites.

A detailed description of the volcanic rocks, associated intrusions and their petrography was given in the second edition of this memoir (Clough and others, 1910, pp. 67–133).

OUTCROP

The main outcrop of the Garleton Hills Volcanic Rocks occupies a 100-km² rectangular area in the north-east part of the district, between the 8-km long, North Berwick coastal section from Canty Bay to Weaklaw Rocks, south to a line through the Alderston Fault, Haddington and Hailes (Figure 17). South of these, the volcanic sequence is thinner and the outcrop is narrow, both north of the Dunbar–Gifford Fault and where the outcrop is detached between that fault and the Lammermuir Fault. To the east the junction with the underlying sediments is complex, due to gentle folding along NE–SW axes. The westerly regional dip takes the volcanic rocks under the overlying sediments, and the concealed extension to the south-west was proved in the IGS Spilmersford Borehole near Pencaitland. Lagios (1984) considered the maximum thickness of the volcanic rocks, on geophysical evidence, to lie under Aberlady Bay.

The basaltic formations are well exposed along the North Berwick coast, both east of the town on the north-west limb of the Balgone Anticline and to the west where the rocks strike W–E until truncated by the Archerfield Fault. A feature of this coast is the number of tuff- and agglomerate-filled volcanic vents. Inland the main basaltic outcrops lie in the Whitekirk and Prestonkirk synclines, producing the rocky hills in the Congalton–Balgone–Stonelaws and Hailes areas respectively. The trachytic rocks, thin and poorly exposed on the coast near Marine Villa, are best exposed where they form rocky hills in the Dirleton–Fenton Barns, the Chesters–Kilduff and the Bangly Hill–Garleton Hills areas. Further south exposures are mostly poor except for the basaltic sections at Blaikie Heugh and the trachytes at Whitelaw. Contemporaneous or later intrusions of many forms are common (Figure 17).

STRATIGRAPHY

The generally good exposure and the large variety of rock types have enabled detailed mapping of the volcanic rocks across the district, and the subdivision into members (Table 1). The members may interdigitate and have diachronous boundaries, but the general stratigraphy persists across the district (Figure 18).

The basal basaltic tuffs, which reach their maximum thickness and are well exposed along the coast east of North Berwick, are called the North Berwick Member, in preference to the term Oxroad Group used by Kelling (*in* Mitchell and others, 1960). Contemporaneous with these are tuff- and agglomerate-filled vents. The overlying basaltic lavas are subdivided into two members. The East Linton Member is characterised by augite-phyric flows such as olivine-basalts of Craiglockhart and Dunsapie types (ankaramites) and interbedded trachybasalts such as kulaites and mugearites. The Hailes Member is dominantly composed of feldspar-phyric olivine-basalts of Markle type (hawaiites) with mugearites. Although generally distinct there is a gradual change between the two members in places. The thickest developments of the basaltic members are preserved in the Whitekirk Syncline and around Hailes. In the west the uppermost group, the Bangley Member, comprising trachytic lavas and tuffs, is thickest around Bangley Hill, the Garleton Hills, Kilduff Hill and Chesters, thinning to north, west and south.

Table 1 Members in the Garleton Hills Volcanic Rocks

Bangley Member	Trachyte, quartz-trachyte and quartz-banakite lavas, trachytic tuffs (0 to 160 m)
Hailes Member	Markle basalts and mugearites (25 to 70 m)
East Linton Member	Mostly Dunsapie and Craiglockhart basalts, mugearites and kulaites (10 to 90 m)
North Berwick Member	Red basaltic tuffs and agglomerates, green basaltic tuffs and agglomerates, freshwater limestones and cementstone ribs (50 to 150 m)

EARLY VOLCANICITY

The earliest evidence of volcanic activity during Carboniferous times in the district is the thin bed of volcanic rocks taken to mark the base of the Calciferous Sandstone Measures at Humbie Mill. A 5-m thick band of purple, tuffaceous sandstones, tuffs and agglomeratic tuffs was recorded approximately 110 m below the Garleton Hills Volcanic Rocks in the Spilmersford Borehole (Davies, 1974, p. 7).

NORTH BERWICK MEMBER

The thickest development of this basal pyroclastic formation is around North Berwick. Martin (1955) estimated the thickness at 220 m. He recognised a threefold division into a lower group of green tuffs, an intervening limestone or cementstone horizon, and an upper group of red tuffs. This subdivision does not hold far south of North Berwick, where the tuffs are generally all red. The calcareous horizon is probably equivalent to the Sunnyside Limestone of East Linton (Bailey *in* Clough and others, 1910, pp. 68, 197).

The member thins southwards, though exposure is poor round the Balgone, Crauchie and Traprain anticlines and in the faulted segment near Gifford. To the south-west at Spilmersford the pyroclastic rocks are 67 m thick, divisible into red beds above and green beds below, but with no calcareous strata (Davies, 1974, p. 21).

DETAILS

North Berwick

The green beds at the base of the sequence were estimated by Martin (1955) to reach 120 m in thickness in the North Berwick area.

Excellent cliff and foreshore exposures occur east of North Berwick from the Yellow Craig Plug to Canty Bay. Green, well-bedded, waterlain, coarse- to fine-grained tuffs are interbedded with tuffaceous sediments and cementstone ribs, all commonly cut by carbonate veins (Martin, 1955; Kelling *in* Mitchell and others, 1960; McAdam *in* Craig and Duff, 1975). Green tuffs are also exposed in the dome-shaped inlier on the shore at Point Garry, west of North Berwick.

The calcareous horizon crops out across the foreshore near the Yellow Craig Plug, east of North Berwick, where 3 m of grey mudstones interbedded with cementstone ribs lie between the green tuffs and the red tuffs. This carbonate horizon is repeated round the inlier of green tuffs at Point Garry. The Rhodes Limestone, formerly seen in Rhodes Quarry [571 851], may have been on the same horizon; it was 9 m of unfossiliferous, blue-grey, massive, compact limestone (Maufe and Bailey *in* Clough and others, 1910, pp. 66–67).

The foreshore between North Berwick Harbour and Yellow Craig has excellent exposures of red, bedded tuffs and tuffaceous marls (Plate 9). They are brick-red in colour, well bedded indicating deposition in shallow water, and vary from fine-grained marly bands to coarse tuffs with larger lapilli, bombs and pyroclastic fragments (Martin, 1955; Kelling *in* Mitchell and others, 1960; McAdam *in* Craig and Duff, 1975). Other outcrops of red beds, repeated by faulting, occur on the foreshore west of North Berwick towards Cowton Rocks. Day (1928b) noted silicified tuffs at Eel Burn.

Balgone Anticline

The outcrop of the member round the anticline is little known. Calcareous beds occur to the east of North Berwick Law, and on the south limb of the anticline at Carperstane, north of Balgone and at Blackdykes north-east of Balgone (Clough and others, 1910, p. 67). It is on the basis of these that green beds are mapped, but they probably die out southwards leaving only red beds.

Crauchie Anticline and Traprain Anticline

The member is mapped mainly on soil colour. The beds thin markedly, particularly around Traprain, and only red beds are present. The Sunnyside Limestone (Clough and others, 1910, p. 68) is developed in the area north-east of Blaikie Heugh but is better known to the east of the present district.

Gifford

In this area little is known of the basal pyroclastic rocks, which are intruded by the Limplum Sill.

Spilmersford

In the Spilmersford Borehole (Davies, 1974) the lowest 30 m of tuffaceous rocks were dominantly green in colour, being green-grey, fine- to coarse-grained tuffs, with thin agglomerate bands and purple-stained in parts. The overlying red beds were 37 m thick, the lower 12.5 m being mainly agglomeratic and the upper 24.5 m mainly tuffaceous. In colour the beds are variously purple and greenish purple, and are poorly bedded, with grain-size variation due more to air-fall than water-sorting. The fragments are mainly of fine basaltic lava, with more rarely sediment and tuff fragments.

Key to letters denoting localities in Figure 17 (opposite)
Necks or vents
EV - Eyebroughy, GV - Garleton Hills, HV - Horseshoe, PV - Partan Craig, QV - Quarrel Sands, RV - Redside, WV - Weaklaw, YV - Yellow Craig Plantation, ZV - Yellow Man.
Cryptovents
GC - Gosford Bay, KC - Kilspindie, PC - Port Seton beach, TC - The Lecks, YC - Yellow Craig.
Trachytic intrusions
At - Athelstaneford Sill, Ba - Bangley Dyke, Bu -Burnside ?sill, Ca - Carperstane Sill, Ey - Eyebroughy Sill, Ga - Garvald laccolith/sill, Ha - Hairy Craig Sill, Ma -Markel ?plug, No - North Berwick Law Plug, Pe - Pencraig laccolith/sill, Tr - Traprain Law Laccolith.
Basaltic intrusions
BM - Burning Mount Plug, BO - Bonnington ?sill, BR - Black Rocks Sill, CA - Cairndinnis Plug, CB - Cheese Bay Sill, CQ - Chesters Quarry Sill, CR - Craigleith Laccolith, DI - Dirleton plugs, FI - Fidra Sill, GB - Gosford Bay Sill, LA - The Lamb Sill, LE - The Leithies Sill, LG - Lower Gullane Head Sill, LI - Limplum Sill, LO - Longskelly Sill, MM - Markle Mains ?plugs, NB - North Berwick Abbey ?plug, NM - New Mains ?sill, PG - Point Garry Sill, PL - Prestonlinks offshore dyke, PR - Prestonpans–Seton Dyke, PS - Port Seton–Spittal Dyke, SP - Spilmersford intrusion, UG - Upper Gullane Head Sill, WF - West Fenton ?sill, YC - Yellow Craig plug/dyke, YP - Yellow Craig Plantation Plug, ZV -Yellow Man dykes.

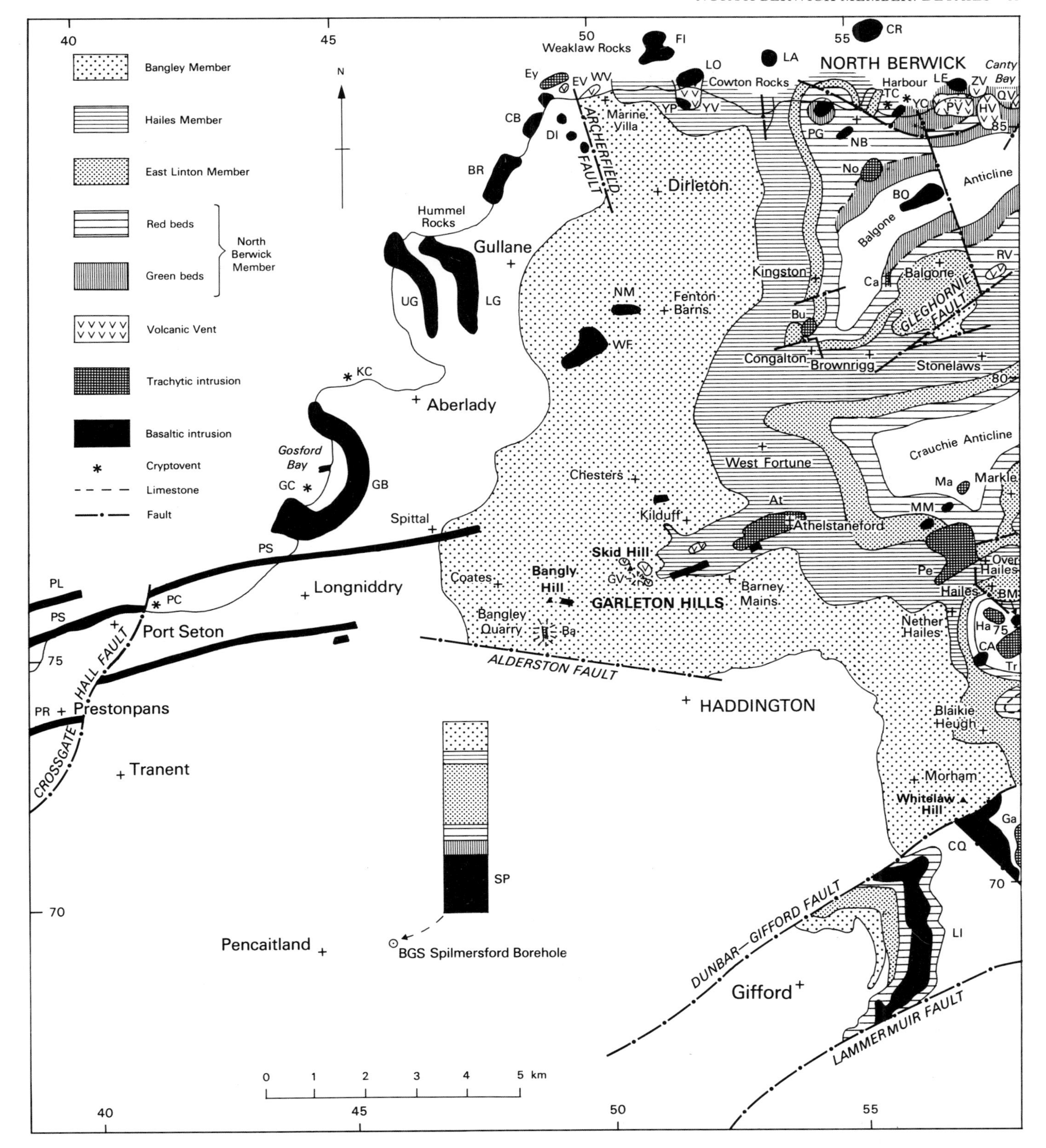

Figure 17 Map of the outcrops of the Garleton Hills Volcanic Rocks and of the vents and intrusions in the district

Plate 9 Red, bedded tuffs overlain by basalt lavas, North Berwick; kulaite lava in middle, cliff of Dunsapie basalt lava behind (D3042)

EAST LINTON MEMBER

The lowest group of lavas in the Garleton Hills Volcanic Rocks, the East Linton Member, is mainly of macroporphyritic olivine-basalts either of Craiglockhart type or Dunsapie type and characterised by abundant, large, augite phenocrysts. The earliest flow or flows are usually non-porphyritic trachybasalts, either mugearite or the unusual leucite-bearing kulaite. Normally these are succeeded by one or more flows of Craiglockhart-type basalt and several flows of Dunsapie-type basalt. The formation is best developed and the lava flows are most numerous in the east around Sheriffhall and Markle where there are about nine flows and the member reaches its maximum thickness at about 90 m. The member is reduced to two flows on the well-exposed section at North Berwick, and is probably represented by only one or two flows in the detached outcrop between the Dunbar– Gifford and Southern Upland faults around Gifford (Figure 17). At Spilmersford the distinction between the East Linton and the overlying Hailes Member is not clear as basalts of Markle-type interdigitate with those of Dunsapie type. Basaltic lavas are commonly 5 to 15 m thick, with as much as half the flow comprising the soft slaggy vesicular top part (Plate 10). Typical lava trap-featuring produced by selective weathering of the hard and soft parts of the flows is normally well displayed even where affected by glacial grooving at right angles to the strike.

DETAILS

North Berwick

Well exposed at North Berwick harbour the member is reduced to two lava flows (McAdam *in* Craig and Duff, 1975). The first lava, about 4 m thick, resting on red tuffs and tuffaceous marls, is an unusual trachybasalt, classed as a leucite-kulaite (Plate 9). Much of the flow is the purple, amygdaloidal or vesicular, autobrecciated top part of the lava. A 30-cm band of massive, fine-grained, non-porphyritic, platy trachybasalt lies above a reddened basal few cen-

Figure 18 Comparative sections through the Garleton Hills Volcanic Rocks.

1 Section recorded in the IGS Spilmersford Borehole [4570 6902].

2 Composite section from field evidence in the Garleton Hills from Bangley [48 75] to Markle [58 77], the section including the five basal flows and the underlying rocks was recorded in the IGS East Linton Borehole [5966 7709] in the adjacent Dunbar district.

3 Composite section from the North Berwick coastal section between Weaklaw Rocks [50 86] and Canty Bay [58 85].

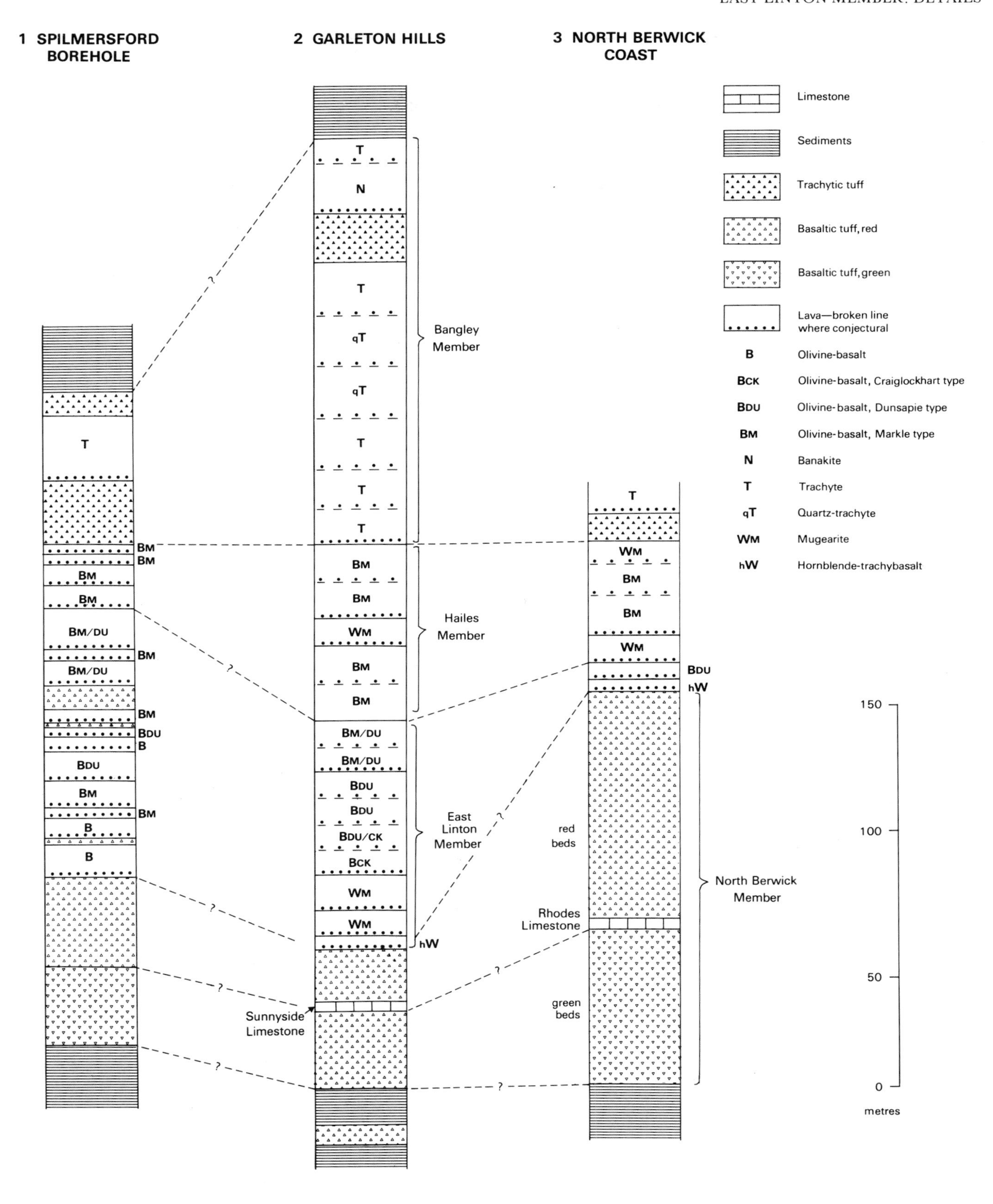
1 SPILMERSFORD BOREHOLE
2 GARLETON HILLS
3 NORTH BERWICK COAST
Limestone
Sediments
Trachytic tuff
Basaltic tuff, red
Basaltic tuff, green
Lava—broken line where conjectural
B Olivine-basalt
BCK Olivine-basalt, Craiglockhart type
BDU Olivine-basalt, Dunsapie type
BM Olivine-basalt, Markle type
N Banakite
T Trachyte
qT Quartz-trachyte
WM Mugearite
hW Hornblende-trachybasalt
Bangley Member
Hailes Member
East Linton Member
North Berwick Member
Sunnyside Limestone
Rhodes Limestone
red beds
green beds
150
100
50
0
metres

timetres incorporating some of the tuff. The rock is highly altered.

Separated by about 1 m of red tuff is a massive olivine-basalt flow, 7 m thick (Plate 9), not 24 to 30 m as suggested by Martin (1955), which would necessitate a near vertical dip rather than the 15° dip observed in the tuffs. The basalt is dark grey to purple-weathered, fine-grained, massive with large augite phenocrysts, red-brown altered olivine phenocrysts and a few glassy feldspars, in a typical basaltic matrix. The rock is intermediate in type between Dunsapie and Craiglockhart types. Rare gabbroic xenoliths a few centimetres across are generally weathered out (McAdam and others *in* Upton, 1969).

To the west along the coast, although affected by faulting and folding, the Dunsapie basalt lava can be traced through the offshore islets of Leap Rock, Craiganteugh, and the Hummell Ridges and at the east end of Cowton Rocks where the dip is to the west. There is no evidence of the kulaite lava in the area, though comparable rocks can be traced to the south for several kilometres. Inland there are no exposures of these rocks till Kingston.

Kingston to Congalton

In this area exposures of the East Linton Member are generally poor, but indications are that the sequence is similar to that at North Berwick. Outcrops show at least one flow of dark grey olivine-basalt with sparse phenocrysts similar to Dunsapie type, though outcrops of a microporphyritic basalt with affinities to Dalmeny type are also present. There is no evidence of the leucite-kulaite.

Congalton to Balgone

A large area of basalt outcrop lies between the Gleghornie Fault and the scarp cliff of Rockville Heughs and Balgone Heughs which run between Congalton and Balgone. Much of the outcrop is a dip slope in a single flow on the NW limb of the Whitekirk Syncline. The base of the formation is poorly exposed. In one place 1 m of tuff is exposed below an olivine-basalt lava, but there is no evidence of the leucite-kulaite. At Rockville Heughs one flow of Dunsapie basalt lies below a sparsely feldspar-phyric mugearite. At Balgone Heughs the 15-m cliff face consists of Dunsapie basalt, possibly formed from a single flow. There is evidence of more than one flow at the west end of Balgone Heughs where 3 m of the vesicular top of one flow is overlain by a 12-m thick Dunsapie basalt lava. Typically the Dunsapie basalt is a grey or purple olivine-basalt with large augite, feldspar and small olivine phenocrysts in a basaltic matrix.

Rocks of the East Linton Member, in the form of olivine-basalts of Dunsapie type, are presumed to continue round the Crauchie Anticline, but are completely covered by drift south from the Gleghornie Fault to the outcrops in the Prestonkirk Syncline at Markle.

Markle

On the north side of the Prestonkirk Syncline at Markle are extensive outcrops in cliffs and small quarries of lavas of the East Linton Member. Exact relationships are complicated by several small faults and by glaciation of the terrain. At least four different lithologies occur, representing four or more lava flows. The lowest flow is a grey or purple, non-porphyritic, platy mugearite. Succeeding this are three flows of olivine-basalt, the lowest flow is a Dunsapie type with phenocrysts of plagioclase, augite and olivine. The middle flow is more basic rock, intermediate between Dunsapie and Craiglockhart types, with numerous augite and olivine phenocrysts and sparse plagioclase. The upper flow is intermediate between Markle and Dunsapie in type with numerous plagioclase, small olivines and a few augite phenocrysts.

On the south side of the syncline a similar succession can be observed. There is an extensive outcrop of lava intermediate be tween Markle and Dunsapie types but the lavas of Dunsapie and Craiglockhart type are poorly exposed. There is no evidence of mugearite or kulaite lavas in this area.

Blaikie Heugh

The East Linton Member thins rapidly to the south, round the Traprain Anticline. Only two types of lava are present at Blaikie Heugh. The basal contact with basaltic tuffs is not exposed. The lower flow is a leucite-kulaite, and the upper one an olivine-basalt of Craiglockhart type. Bennett (1945) described in detail the lower flow as a purple, fine-grained trachybasalt, called a leucite-kulaite, though Bailey (*in* Clough and others, 1910, p. 125) observed a band of cementstone between what he took as two lava flows. The rock has large hornblende pseudomorphs, augite phenocrysts and a few olivine pseudomorphs. The upper flow of olivine-basalt of Craiglockhart type is dark grey with large black augite and red-brown olivine phenocrysts. This flow is in turn overlain by a trachyte lava of the Bangley Member, overstepping the Hailes Member.

Gifford

In the area between the Dunbar–Gifford and the Southern Upland faults, lavas of the East Linton Member die out. The only lava outcropping here is an olivine-basalt of Craiglockhart type.

Spilmersford

The succession proved in the Spilmersford Borehole is lithologically quite unlike the succession in the Garleton Hills area in as much as the clear distinction between augite-phyric olivine-basalts and augite-free rocks is confused by the alternation of olivine basalts of Markle and Dunsapie type, as well as non-porphyritic types (McAdam, 1974). The lower 11 flows from the borehole can be taken as equivalent to the East Linton Member. In these olivine-basalts, four are of Markle type, two intermediate between Markle and Dunsapie types, two of Dunsapie type and three generally non-porphyritic. The thickness of the lavas varies between 4 and 16 m, which bears comparison with the revised thickness of similar lavas at North Berwick, as against the exaggerated thicknesses given in earlier publications (Clough and others, 1910; Martin, 1955).

HAILES MEMBER

These higher flows, probably six at most and up to 70 m thick in total, are characterised by feldspar-phyric basalts of Markle type, including, just east of the district, the type locality of Markle Quarry. There are also sparsely porphyritic to non-porphyritic basalts and mugearites. In the district these rocks are exposed along the coast from North Berwick to Marine Villa. Inland the outcrop forms a strip southwards dying out west of Traprain Law with large areas of well-exposed outcrop caused by the Balgone and Prestonkirk anticlines. Good exposures are seen at the Cogtail Burn valley in the Garleton Hills west of Athelstaneford (Figure 17). Although these rocks form some good scarps and trap-featuring, glaciation has commonly reduced exposures to a complex pattern of isolated knolls.

DETAILS

North Berwick

Two flows are represented in the section at North Berwick harbour, a lower 10-m thick flow of mugearite and an upper olivine-basalt flow of Markle type, some 17 m thick (McAdam *in* Craig and Duff, 1975). The mugearite flow is pale purple-grey, fine-grained, non-porphyritic and platy with iron-banding and carbonate veins. The top third of the flow is auto-brecciated into blocks in which differential weathering picks out an apparent pillow structure, and flow has produced elongate amygdales (Plate 10). The overlying Markle basalt is dark grey to purple with numerous large labradorite phenocrysts and small brown iddingsite pseudomorphs after olivine, in a fine-grained basaltic matrix. This flow also has a distinct slaggy vesicular top with large calcite-filled vugs.

Cowton Rock to Marine Villa

Folding and faulting causes the same rocks to outcrop along the shore further west of North Berwick. At the east end of Cowton Rocks a thin flow of Markle basalt underlies brown-purple, fine-grained, non-porphyritic mugearite, which is possibly in two flows. Because of the low westerly dip the overlying Markle basalt with abundant large white feldspar phenocrysts and small brown olivines crops out extensively at Cowton Rocks and at Longskelly Rocks, possibly representing more than one flow. The Markle basalt is overlain south of Cowton Rocks by a faulted patch of red silicified trachytic tuffs. Towards Marine Villa, however, the Markle basalt is overlain by a flow of pale purple, fine-grained, sparsely to non-porphyritic mugearite, which in turn is overlain by the bedded trachytic tuffs.

Kingston to Congalton to Brownrigg

In this inland area two prominent lava crags run from Kingston to Congalton, the rocks dipping to the west-north-west, and from Congalton to Brownrigg, where the dip is southwards. The section here appears similar to that on the coast with at least four flows present, two flows of Markle basalt alternating with two of mugearite. The lower mugearite is very well exposed in the 7-m face of the quarry south of Kingston in the crags running east from Congalton, hence its provisional name 'Congalton Rock' (Bailey *in* Clough and others, 1910, p. 74). Rare exposures of the contact between the two

Plate 10 Scoriaceous lava-top and flow-structures in mugearite lava, North Berwick (D3041)

lower lavas are seen along the face in the quarry at Kingston. A feature of the purple, fine-grained, sparsely porphyritic mugearite is the concentric iron-banding present.

Stonelaws

South of the Gleghornie Fault, lavas of the Hailes Member lie in the gentle basin structure of the Whitekirk Syncline. The complex pattern of isolated outcrops has been produced by faulting and by glaciation, and nowhere are junctions between the flows visible. Evidence here indicates at least five different lithologies, with alternating Markle basalts and mugearites as seen further west, with at least one additional flow of a Markle type basalt with small feldspar phenocrysts. The full sequence in ascending order is Markle basalt, mugearite, Markle basalt with small phenocrysts, Markle basalt, sparsely porphyritic mugearite or basalt. This is overlain by red trachytic tuffs in a small inlier.

Garleton Hills

Between West Fortune and Athelstaneford are isolated outcrops of a mugearite lava and of an overlying Markle basalt lava, but these are the only evidence of the Hailes Member which is generally drift-covered round the west side of the Crauchie Anticline. In the Garleton Hills higher rocks are seen lying immediately below the trachyte lavas round the glacially-deepened valley of the Cogtail Burn. The lowest rocks exposed are Markle basalts with abundant, large feldspar phenocrysts, in the ravine south of Athelstaneford. On the slopes to the west the rocks are similar Markle basalts with, below Barney Mains, an intervening flow of mugearite which dies out to the west.

Hailes

Much of the high ground east of Over Hailes is underlain by flat-lying lavas, the two lowest of the Hailes Member. These are the same lavas as those exposed in Markle Quarry, the type Markle basalt of Markle Quarry overlain by a sparsely porphyritic basalt related to Markle type. To the south-west around Nether Hailes, higher flows occur, as in Brown Knowe Plantation where a thick mugearite flow appears to lie between two lavas of Markle basalt, overlain in turn by trachyte lavas.

Spilmersford

The four upper flows of the succession proved in the Spilmersford Borehole are all Markle basalt with thicknesses between 2.8 and 9.0 m, each showing a thick vesicular top (McAdam, 1974). There are no mugearites in the succession here.

BANGLEY MEMBER

There is a sharp change from the basaltic effusives that form the lower part of the Garleton Hills Volcanic Rocks to the overlying trachytic lavas and tuffs which comprise the Bangley Member. Unlike the basaltic rocks there is no clear division between the pyroclastic episode and the eruptive episode, trachytic tuffs being interspersed at various horizons within the lavas. The trachyte lavas are less widespread than the basalts. The maximum thickness, estimated at 160 m, occurs in the Garleton Hills with the member thinning out rapidly to the coast in the north, to Gifford in the south and being represented by only one flow to the south-west at Spilmersford (Figure 17). The main area of outcrop lies between Dirleton in the north and the Garleton Hills in the south. The outcrop continues as a narrow strip from east of Haddington to Whitelaw Hill, and between the Dunbar–Gifford Fault and the Southern Upland Fault as a detached outcrop. The trachyte lavas are generally thick, greater than 20 m, though junctions are not seen at outcrop. The one flow proved in the borehole at Spilmersford was over 25 m thick. Probably included within the rocks mapped as lavas are intrusive bodies which have similar lithologies to the lavas (Barrow *in* Clough and others, 1910, p. 75). The rocks include both porphyritic, sparsely porphyritic and non-porphyritic types, with some of the first category containing essential quartz.

DETAILS

Cowton rocks

A faulted outlier of silicified trachytic tuff and conglomeratic tuff lies on top of basalt lavas on the south side of Cowton Rocks (Day, 1928b).

Marine Villa

Overlying the mugearite lava at Weaklaw Rocks there is a red, bedded tuff showing alternation of coarse and fine bands, and an agglomeratic base. Exposures in the cliff at Hanging Rocks, 400 m W of Marine Villa, show 8 m of purple and cream, bedded, trachytic tuff with agglomeratic bands, overlain by a purple, vesicular, porphyritic, trachyte lava, 5 m thick (McAdam *in* Craig and Duff, 1975, p. 90). Trachyte lavas higher in the sequence form crags in the forests inland. Across the Archerfield Fault to the west, strata at a higher level contain tuffaceous dolomitic rocks indicating the presence of pyroclastic rocks higher in the sequence. Rock-bit samples from a water bore at Archerfield indicated 34 m of trachytic lava and marly clay (probably tuff) underlying a white sandstone.

Dirleton to Fenton

North of Stonelaws an outlier of red trachytic tuffs overlies the highest basaltic rocks, mugearite lava. Further west, trachyte lavas probably rest directly on basaltic lavas. Although outcrops are isolated in the area from Dirleton to Fenton Barns there is evidence of several lava flows separated by a thick band of trachytic tuff. Of the three flows below the tuff, the middle lava consists of quartz-trachyte which is seen east of Fenton Barns in an old quarry at East Fenton and in Rattlebags Quarry. The other flows are feldspar-phyric trachyte lavas, the lower being exposed in a large railway cutting whereas the upper one is poorly exposed. Above the tuff lies the thick quartz-trachyte with conspicuous feldspar phenocrysts seen in Craigs Quarry, west of Dirleton. Higher lavas occur further west overlain by till, temporary pipeline exposures showing a thick amygdaloidal top to flat-lying lavas.

Chesters to Kilduff

The hills at Chesters and Kilduff are formed of trachyte lavas. These are similar to the rocks described to the north, and mainly consist of quartz-trachyte. In this area the quartz-trachyte lavas apparently rest directly on the basaltic lavas. There is some evidence of higher feldspar-phyric trachyte lavas to the west, though the area is mostly drift-covered.

Bangley to Garleton Hills

Here the topographic form of the trachytes is partly dependant on the strike of the rocks. Where the strike is W – E and parallel to the direction of glaciation, as in the eastern Garleton Hills, trap featuring is developed as seen at Kae Heughs and Hanging Craig (Frontispiece). Where the strike is N – S and at right angles to the glaciation, as in the western Garleton Hills, there are glacially rounded hills, for example Byres Hill, Skid Hill, Phantassie Hill and Alderston Hill. The thicknesses of the lavas are uncertain as flow-junctions are rarely seen. Some unrecognised intrusive trachytes may also be present. The trachyte flows are normally at least 20 m thick, as the main quarries show single flat-lying flows of this thickness. The stratigraphic sequence of the lavas is not clear. The main variation in lithology appears to be geographical rather than stratigraphical, the eastern and western parts of the Garleton Hills having appreciably different types of trachytes.

In the western Garleton Hills, west of the Mutton Hole Fault, trachytes are generally porphyritic with abundant, large, feldspar phenocrysts. The lowest flow, resting directly on Markle basalt lava, forms Craigy Hill and the rocks immediately to the south and north. The next flow forms the main mass of Skid Hill and is well exposed in the quarry there. The trachyte is a blue-grey, fine-grained rock with glassy sanidine phenocrysts, and has iron-veining and iron-staining varying with the frequency of the joints. A higher flow of similar rock forms Byres Hill on which the Hopetoun Monument stands, Score Hill and probably Phantassie Hill, though the last may be equivalent to the flow at Skid Hill. The rocks forming Pains Hill and the eastern part of Alderston Hill are probably a still higher flow. Above these trachyte lavas lies a thick bed of trachytic tuff, which forms the middle part of Alderston Hill and was formerly well exposed in the quarries at Bangly Braehead where a fine-grained, purple tuff with trachytic fragments was exposed. Above the tuffs lies a thick flow of quartz-banakite which forms Silver Hill, the western part of Alderston Hill and Bangly Hill. This flow is well exposed in the working Bangley Quarry, where the trachyte is a blue-grey rock with large glassy feldspar phenocrysts zoned from sodic plagioclase cores to alkali feldspar rims. The freshness of the rock depends on the frequency of the joints in the quarry, and much of the trachyte is highly weathered. It is possible that one or more of the higher flows lies in the poorly exposed ground around Coats.

The eastern part of the Garleton Hills, lying east of the Mutton Hole Fault, consists of sparsely porphyritic trachyte. The purple trachyte has a granular, platy appearance in hand specimen, with iron-ribbing. There may be at least three flows present, the lower two forming the two scarps at Kae Heughs and along to Hanging Craig and Seaton Law Quarry. A higher flow of the same lithology forms Yellow Craigs and the extensive dip-slope to the south. The highest volcanic rocks in the area are trachytic tuffs, formerly well exposed in the infilled Abbey Quarry, where coarse-grained, brown agglomeratic tuff with trachytic fragments was recorded.

Whitelaw Hill

South of the River Tyne, scattered outcrops of trachytic rocks occur from Hailes to Morham, but the main outcrop is at Whitelaw Hill. Various trachyte lithologies are present, but the relationship between them is not clear. Trachytic tuffs have been mapped within the lavas west of Nether Hailes and at Morham Church, and above the lavas in the area of the Bearford Burn.

Gifford

Trachyte lavas occur between the Dunbar – Gifford Fault and the Lammermuir Fault north-east of Gifford. The exposure in this area is poor and little is known about the trachytes.

Spilmersford

In the Spilmersford Borehole section the Bangley Member consists of three parts. Basal trachytic tuffs and agglomerates, 25 m thick, lie directly on basalt lavas. In the middle is the only trachytic lava, a porphyritic trachybasalt, over 25 m thick, and this is overlain by 11 m of pyroclastic rocks. The tuffs below the trachybasalt lava are purple, fine-grained, unbedded, with a few larger fragments, whereas the agglomerates are generally greenish with non-porphyritic trachytic fragments up to 1 cm. The trachybasalt lava is green-grey, fine-grained, with numerous large feldspar phenocrysts which consist of single crystals or glomeroporphyritic aggregates of potash feldspar and calcic plagioclase. There are also microphenocrysts of altered olivine. This rock has features which differ from the lavas seen in the Garleton Hills. The upper tuffs and agglomerates are mainly purple or chocolate-purple with fragments up to 10 cm of porphyritic trachyte.

VOLCANIC VENTS

Along the north coast from Eyebroughy to Canty Bay there are four large and several small, mainly agglomerate-filled volcanic vents. Most of these vents were first recognised by Day (1923; 1925b; 1928a; 1932c). The larger vents are the Yellow Craigs Plantation Vent, the Partan Craig Vent, the Yellow Man Vent and the Horsehoe Vent. The smaller coastal vents are at Eyebroughy, Weaklaw and Quarrel Sands. Inland there is a complex vent near Redside and four small vents in the Garleton Hills. Cryptovents have been described from several coastal localities from Port Seton to Aberlady and at North Berwick (Figure 17).

Martin (1955) divided the vents into a younger Green Group which includes the four larger vents and an older Red Group represented in this district by a small vent at Quarrel Sands. Martin showed that the Red Group vents were active prior to the eruption of the lavas. Green Group vents, on the other hand, cut strata up to the Hailes Member and contain fragments of Markle basalts and mugearites.

DETAILS (letters refer to location map Figure 17)

Vents of the Red Group

Quarrel Sands Vent (QV)

The only vent of this group in the district is poorly exposed on the foreshore at Quarrel Sands, where red volcanic tuff with bombs cuts the sedimentary rocks well below the Garleton Hills Volcanic Rocks.

Vents of the Green Group

Yellow Craig Plantation Vent (YV)

Martin (1955) classed as one structure, two vents described by Day (1932b) as the Longskelly Rocks Vent and the Yellow Craig Plantation Vent. The north part of this vent is exposed on the foreshore at Longskelly Point, where a western contact is visible with Markle basalt lava of the Hailes Member. The vent is infilled by green tuff with numerous basaltic bombs and blocks. Intruded into this tuff is the Longskelly Sill. The south part of the vent, though mainly covered by raised beach and blown sand has the conical hill of the Yellow Craig Plantation Plug. Exposures of the vent agglomerate north-east of the plug show yellow-brown, fine-grained tuff with blocks of Markle basalt lava, mugearite, bedded tuff, sandstone and siltstone (McAdam *in* Craig and Duff, 1975, p. 86). As it cuts the Hailes Member lavas and contains fragments of these lavas, the vent is clearly later than these rocks.

Partan Craig Vent (PV)

This interesting and well-exposed vent lies just east of North Berwick. Day (1925b) first recorded this vent which Martin (1955) considered to be a double vent. Contacts with green, bedded tuff of the North Berwick Member at the west and north-east sides are exposed on the foreshore, whereas the other contacts are obscured or under water. The vent is entirely filled with agglomerate and tuff, with no evidence of intrusive bodies. The west part of the vent shows a collapse-synclinal structure, exposed in the west-facing cliff of Partan Craig, and in the basin structure to the north. The vent-rock in the west consists of poorly bedded, reddish-green, agglomeratic tuff, containing large blocks of red-green, bedded tuff and tuffaceous sandstone, red siltstone and pale cementstone, with bombs of nepheline-basanite, a rock-type found locally only in intrusions. At the south-west end of Partan Craig is an area of particularly massive blocks of a tuff, possibly a debris-flow (Plate 11). The east part of the vent contains green tuff with predominantly igneous bombs. Thin, calcite veins are a feature of the vent. To the west of the vent is an area of disturbed tuff, from which has been recorded blocks of metamorphic rock comparable with Precambrian rocks, brought up from the rocks lying under the Upper Palaeozoic cover of the Midland Valley (Graham and Upton, 1978; Upton and others, 1983).

Yellow Man Vent (ZV)

Sandwiched between its two larger neighbours, the Partan Craig and Horseshoe vents, is the Yellow Man Vent, first described in detail by Day (1925b). The vent cuts green, bedded basaltic tuff of the North Berwick Member, contacts with which are exposed on the foreshore. The vent aggomerate comprises green or reddish green, coarse-grained tuff, containing large blocks up to 3 m across of bedded red and green tuff, large and small blocks and bombs of basanite, and fragments from lower down the succession, for example red sandstone, cornstone and lava of Old Red Sandstone

Plate 11 Vent – agglomerate or debris-flow, Partan Craig Vent (K3046)

aspect, and cementstone. A feature of this vent is the associated intrusions within and just outside the vent which take the form of small plugs and irregular dykes of basanite.

Horseshoe Vent (HV)

Just to the east of the Yellow Man Vent lies another large vent filled with green tuff and agglomerate, intruded into green, bedded, basaltic tuffs of the same member. The similarity of the vent and country rocks masks the distinction, but the detailed outline of the vent given by Day (1928a) is generally demonstrable. The north and east margins of the vent can be traced across the intertidal rock platform of Leckmoran Ness from the Yellow Man Vent, which cuts the Horseshoe Vent, to Horseshoe Point. The vent-agglomerate consists of green, poorly bedded tuff containing basanite bombs, in places numerous, and blocks of bedded tuff, sandstone and mudstone.

Other Vents

Eyebroughy Vent (EV)

A poorly exposed intertidal area of yellow breccia with large blocks of fine-grained igneous rocks, between Eyebroughy and the coast, is taken as a vent.

Weaklaw Vent (WV)

Day (1923) identified and described in great detail an area of Weaklaw Rocks as a volcanic vent cutting a pre-existing intrusion-breccia along a fault striking NE–SW. The fault-plane forms the spectacular bluff feature of Hanging Rocks. The intrusive breccia and vent have been shown together on Figure 17. Francis (1960) compared the intrusion breccia to similar features described from Fife.

Redside Vent (RV)

Between Redside and East Craigs are rocky knolls containing within a very small area outcrops of tuff, mugearite, and olivine-basalts of Markle, Dunsapie and Craiglockhart types. An explanation of the juxtaposition of these lithologies is that these mark the site of a vent with a long history of eruption.

Garleton Hills (GV)

In the Garleton Hills are four small areas which have agglomerate, tuff and trachyte breccia, and which have been identified as possible volcanic vents (Figure 17).

CRYPTOVENTS

At several localities along the coast Francis (1960) and Howells (1969) have identified features called cryptovents. Francis (*in* Craig and Duff, 1975) described cryptovents as small ring structures which represent upward drilling of gas-propelled steam advancing above the molten magma column and containing ragged-edged fragments of chilled basalt and of sediments broken by the steam from walls of the channel. The drillings were arrested before reaching the surface.

DETAILS

Gosford Bay Cryptovent (GC)

Howells (1969) described a cryptovent on the shore at Gosford Bay which is an oval-shaped ring fracture cutting Lower Limestone Group sedimentary rocks. Large blocks of flaggy, massive sandstone, showing steep dips into the centre of the ring structure are separated by coarse intrusive breccias and tuffs containing coal and shale clasts occurring in a sandy matrix with basaltic glass. The cryptovent has been linked to the intrusion of the Gosford Sill a short distance below.

Other cryptovents identified by Howells (1969) at Gosford, Port Seton beach (PC) and Kilspindie shore (KC) show the disturbed blocks in a ring structure but the intrusive basaltic material is not present and this is taken to indicate conditions even closer to the contemporaneous surface.

The Lecks Cryptovent (TC)

Cutting the red basaltic tuff high in the North Berwick Member on the foreshore east of North Berwick is a cryptovent structure described in detail by Bennett (1945). Within the ill-defined area of the vent are disordered blocks of red tuff, up to 1 m or more across, in a sparse matrix of red and green agglomeratic tuff. Bennett suggested that the disturbance occurred a short period after deposition of the tuffs. The vent is pierced by four small analcime-basanite intrusions which have baked the surrounding tuff.

Yellow Craig cryptovents (YC)

Just to the east, three small cryptovents occur at Yellow Craig, one round the north end of the Yellow Craig Plug, and two cut by the dykes which run to the north-east. All of these cryptovents contain disorientated blocks of red tuff, green tuff and cementstone, which are pierced by the cryptovents, and it is clear that some material has been moved up the vent and some down (McAdam *in* Craig and Duff, 1975). ADM

PETROGRAPHY

The nomenclature used on the geological map of the district and in this account is the petrographical classification of Carboniferous basalts formulated by MacGregor (1928). This classification divides the basaltic rocks into six main types, with transitions, depending mainly on the type and size of phenocrysts but also on the generally feldspathic or pyroxenic nature of the rock. A more recent classification (Macdonald, 1975) of such rocks was proposed, based on their chemical composition and hence on the normative minerals. Macdonald's correlation between the two classifications is shown, with minor modifications, in Table 2.

The use of a classification based on the petrochemistry of the rocks is more helpful in a study of the geochemistry, provenance and petrogenesis of a volcanic suite. It is, however, less useful for mapping purposes unless large numbers of analyses are available and is not useful in dealing with altered rocks. MacGregor (1928, p. 348) pointed out that rocks similar, in many respects, to mugearites are produced by the more or less complete albitisation of some fine-grained basalts. Apart from the petrographic difficulties, such partial or complete albitisation can strongly affect a chemical classification and may cause an altered basalt to

Table 2 Classification of basalts
Comparison of the basalt nomenclature of MacGregor (1928), based on phenocryst type and size, with the chemical classification of Macdonald (1975) for the mildly alkaline and transitional basic rocks.

	MacGregor (1928)	Phenocrysts Abundant	Phenocrysts May be present	Macdonald (1975)
Microporphyritic (<2 mm)	Jedburgh +	Plag, 01		Microphyric 01-plag ± Fe oxides – phyric (basaltic) hawaiites (occasionally basalt)
	Dalmeny +	01	Cpx, Plag	Microphyric 01 – or 01 + Cpx-phyric basalt
	Hillhouse +	01	Cpx	Microphyric 01 ± Cpx-phyric basalt
Macroporphyritic (>2 mm)	Markle	Plag	01	Macroporphyritic 01 – plag ± Fe oxides – phyric basalts, basaltic hawaiites or hawaiites
	Dunsapie	Plag, 01, Cpx		Macroporphyritic 01 – cpx-plag-Fe oxides – phyric basaltic hawaiites, or 01-cpx-plag-phyric basalts
	Craiglockhart	01, Cpx		Ankaramite

01 = olivine, Cpx = clinopyroxene, Plag = plagioclase, + = Aphyric varieties classed according to matrix composition (proportion of feldspar and pyroxene)

have the necessary composition to be classed as hawaiite or mugearite.

Basalts in the district are mainly of Markle type in the Hailes Member and mainly Dunsapie and Craiglockhart types in the East Linton Member. Varieties transitional be tween the macroporphyritic types have been recorded and microprophyritic Jedburgh and Dalmeny types, and transitions between the microporphyritic and macroporphyritic types, are rare. In some flows, where phenocrysts are less abundant and irregularly distributed, individual specimens may not possess all the attributes of a specific type.

Mugearites occur in both the Hailes and East Linton members but 'leucite' hornblende-kulaite occurs only at the base of the lava sequence.

Trachytic rocks include quartz-trachyandesites (quartz-banakites) and porphyritic and aphyric varieties of trachyte, some with appreciable quartz, form the upper part of the volcanic sequence, the Bangley Member.

DETAILS

Basalts and hawaiites

Olivine-basalts of Craiglockhart type (ankaramites)

The olivine phenocrysts are represented only by pseudomorphs in carbonate, hematite, serpentine and quartz whereas the large phenocrysts of pale purple augite (S 1004, 49353) are only locally altered. The matrix is commonly fine-grained with microlitic augite and reminiscent of that of the fine-grained basanite intrusions, but some (S 11837, 54538) are coarser and more feldspathic (S 11837). Locally nests of finely granular augite occur (S 11329, 11844). Sliced rocks (S 11329, 11844, 50348) from Blaikie Heugh contain large areas of analcime in the base giving rise, when examined under cross polars, to a blotchy effect like that observed in the Kidlaw basanite (p. 68). A similar effect has been observed in a specimen (S 49353) from Hailes area. Transitions to Dunsapie type are represented by rocks (S 11837, 54538) containing rare phenocrysts of plagioclase.

Olivine-basalts of Dunsapie type

Though all basalts of this type are characterised by plentiful phenocrysts of olivine, augite and plagioclase, the high proportion of plagioclase in a number of specimens (S 11340–1, 54548) would assign them to Clark's (1956, p. 46) 'Feldspathic Dunsapie type' or indicate a transition to Markle type. The olivines are normally represented by pseudomorphs and the phenocrysts of pale augite also are commonly altered. The pale augite, or its pseudomorphs, in the groundmass occurs as small granules or grains, or in coarser varieties, as prisms. The matrix may be relatively coarse-grained yet the pyroxene finely microlitic (S 54548). Ophitic augite has not been seen in rocks of this type in the district. Several sliced specimens (S 10583, 51263, 54356) from the Stonelaws–Balgone area are very fine-grained and closely resemble fine-grained varieties of the lava of Kippielaw to the east of the district. In a number of specimens of this variety, biotite of late crystallisation occurs in the groundmass (S 11341, 52084, 53410, 54545) and, in some instances, in vesicles (S 11340, 51258, 51263, 52086, 54356). In addition, rare crystals of brown hornblende, with large optic axial angle, have been noted (S 53410, 54545). Apatite is a common accessory as needles and occurs rarely (S 53211) as large crystals.

Olivine-basalts (and hawaiites) of Markle type

Basalts of Markle type are common in the district. They contain many phenocrysts of zoned calcic plagioclase though they are not as porphyritic as the exceptionally porphyritic rock of the type locality,

Markle Quarry, just beyond the eastern margin of the district. The anorthite content of the calcic cores of the plagioclase is greater than An_{65} and values of up to An_{70} (S 752) and An_{74} (S 51152) have been recorded. The plagioclase may be albitised and/or carbonated. Olivine, generally represented only by pseudomorphs in hematite, goethite, carbonate and serpentine occurs as phenocrysts, microphenocrysts and small crystals in the groundmass. Residual fresh olivine has been noted only rarely (S 11938). Augite, generally very pale, though locally pale purple-brown (S 10588), occurs as small prisms and grains in the matrix. Commonly the augite is replaced by chlorite or carbonate. Iron ore occurs as small grains and octahedra and, rarely, (S 53210) as large plates. The groundmass-feldspar constitutes small laths generally zoned from labradorite to andesine. Oligoclase is common, mantling the more calcic plagioclase and is present in increasing quantity in rocks transitional to mugearite (S 1032, 10820). In a number of samples the plagioclase, both phenocrysts and groundmass, is partly or completely albitised. Albitisation is a particularly noticeable feature in lavas of this type from the Spilmersford Borehole (S 53192 – 96, 53209, 53218 – 19) described by McAdam (1974, p. 41) though residual scraps of calcic plagioclase may still occur (S 53206). Analcime has been observed in two very fresh basalts (S 752, 11938) from the Marine Villa – Cowton Rocks area. Needles of apatite are common and some large crystals occur (S 53209 – 10). Very coarse varieties with grain-size approaching dolerite also occur.

In some specimens (S 10588, 10816, 33579, 51241) the phenocrysts of plagioclase are mainly about 2 mm diameter or are sporadic and, according to MacGregor (1928, p. 350) would be regarded as transitional between Jedburgh and Markle types. One rock (S 52101) of this variety differs in having subophitic to ophitic plates of augite.

Rocks transitional to Dunsapie type include those (S 51231) in which fairly numerous microphenocrysts of pale purple augite occur in addition to phenocrysts of plagioclase and olivine. Some (S 51241) with numerous small phenocrysts (2 mm or less) of plagioclase and many of olivine can be classed as transitional between Dalmeny and Markle types (MacGregor, 1928, p. 350).

Microporphyritic olivine-basalts (and hawaiites)

Basalts of Jedburgh type occur, albeit rarely. One (S 54534) is very coarse-grained and characterised by ophitic plates of largely carbonated purple augite. Rocks of mugearitic aspect, recorded by McAdam (1974, p. 42), in the Spilmersford Borehole contain pseudomorphs after ophitic augite probably represent albitised basalts of Jedburgh type (S 53214 – 15). Two sliced specimens (S 51244 – 45) of altered ophitic basalt contain rare feldspar phenocrysts and may be classed as intermediate between Dalmeny and Markle types. Partly albitised basalt allied to Jedburgh type (S 52102) but probably transitional to mugearite has been noted in the Stonelaws – Balgone area.

Mugearites and trachybasalts

Mugearites

The rocks in this district which have been grouped under the heading mugearite probably cover a range of compositions from hawaiite through mugearite to benmoreite. The Carboniferous lava first noted as resembling Tertiary mugearites was that of Congalton Mains in the Hailes Member (Bailey *in* Clough and others, 1910, pp. 74, 124). Bailey described the rock as containing as much orthoclase as oligoclase. This is not typical of mugearites in general nor of the Scottish Carboniferous mugearites (MacGregor, 1928, p. 347) in particular, in which orthoclase is not an essential constituent. Further work on the Congalton Mains rock by Muir and Tilley (1961, table 4) showed the feldspar to range from anorthoclase to soda-sanidine and they described the rock as mugearite-trachyte. The term benmoreite was later introduced by Tilley and Muir (1964, p. 439) to cover rocks transitional between mugearite and trachyte and this term was applied to the Congalton Mains rock by Macdonald (1975, p. 312). Other rocks (S 10577 – 78, 36474) transitional to trachyte and containing much oligoclase and alkali-feldspar occur at North Berwick.

Some rocks grouped with the mugearites may represent basaltic mugearites or hawaiites. They (S 10831, 50154) contain much labradorite in addition to oligoclase and have small olivine pseudomorphs, many granules of augite and grains or iron ore. A porphyritic variant (S 11348) contains phenocrysts of labradorite (An_{60}) in a matrix of oligoclase laths which have slender cores of labradorite (An_{52}). Similar porphyritic varieties have phenocrysts of calcic plagioclase (S 51234) or albitised plagioclase (S 10831) in a matrix of small laths of labradorite, much oligoclase, small olivine pseudomorphs and granular augite and iron ore. Some (S 50169, 51279) contain phenocrysts of alkali-feldspar.

More typical of the mugearites are rocks (S 10804 – 5), from the Garleton Hills, composed of a fluxioned felt of small poorly shaped oligoclase laths with much iron ore, tiny pseudomorphs after augite and some altered olivine.

The matrix of the rocks commonly has well-developed planar flow structure and the feldspars are flow-oriented slender laths or microlites. More rarely (S 10582, 10835, 51246) however, the matrix-feldspar consists of somewhat orthophyric, rectangular laths.

Biotite is present in a number of sliced rocks (S 50154, 51147, 54627) as poorly shaped flakes and patches, and brown hornblende occurs less commonly (S 10829, 36476). Both minerals also occur locally, as noted elsewhere in Scottish lavas (MacGregor, 1928), in the feldspathic basalts (Jedburgh and Markle types).

'Leucite'-kulaite ('leucite'-hornblende-trachybasalt).

This unusual rock-type has been recorded at the base of the lava sequence (in the East Linton Member) at only two localities in the district—at North Berwick and, in the south, at Blaikie Heugh. As described by Bailey (*in* Clough and others, 1910, pp. 125 – 126) the rock is characterised by a profusion of euhedral hornblende pseudomorphs among its phenocrysts. The hornblende is almost entirely resorbed and only rarely (S 11244, 11328) are kernels of the original brown mineral preserved. The other phenocrysts are euhedral, violet-brown augites with a few pseudomorphs after olivine. Some large prisms of clouded apatite occur. The groundmass contains laths of oligoclase mantled by potash-felspar, small prisms of augite, much iron ore and minute flakes of biotite. Large patches of analcime occur which locally contain concentrically arranged inclusions (S 11244) and Bailey and later Bennett (1945, pp. 38, 40) considered that these represented the alteration of original leucite.

Trachytes and quartz-banakites

As noted by Bailey (*in* Clough and others, 1910, p. 127) the trachytic group of lavas, in this district, is divisible into quartz-banakites (trachyandesites with frequent plagioclase phenocrysts and a little quartz in the groundmass), porphyritic sanidine-trachytes, non-porphyritic trachytes and quartz-trachytes (interstitial quartz more important and including porphyritic and non-porphyritic varieties). Flow structure is generally well developed in the matrix of all types and locally well developed lines of 'ruck' or flow shear are present (S 50146 – 47, 50155, 50157) (Plate 16.6).

Quartz-banakites

These rocks are characterised by the presence of plagioclase phenocrysts commonly mantled by potash-feldspar and in many instances with phenocrysts of potash-feldspar set in a trachytic groundmass of laths or microlites of orthoclase with patches of quartz. Rocks at Bangley Quarry (S 625, 1006, 1690, 30688, 46167–69, 46171), Skid Hill Quarry (S 4517–18, 30687) and Kae Heughs (S 1007–8) are of this type. The Skid Hill rock represents a transition to the porphyritic quartz-trachytes in that the proportion of plagioclase is smaller. The plagioclase is locally (S 46180) pseudomorphed by kaolinite. Pyroxene, usually represented by pseudmorphs, occurs as subhedral to euhedral microphenocrysts and as small pseudomorphs in the matrix. Residual scraps of very pale augite are preserved in altered microphenocrysts in a specimen (S 1683) from Bangley Quarry. A specimen (S 57141) from Skid Hill contains microphenocrysts of augite zoned in colour from pale purple out to pale green and prisms of pale green sodic pyroxene occur in the matrix; some purple augite is enclosed in a plagioclase phenocryst. In this sample yellow-orange isotropic 'chloritic' material occurs intersertally and also as pseudomorphs reminiscent of those after fayalite in the Traprain Law and Bass Rock intrusions. Brown hornblende occurs as a small microphenocryst and as anhedral plates moulded on feldspar in a sliced rock (S 1683) from Bangley Quarry. Biotite phenocrysts occur in two sliced rocks (S 46169, 46180) from the Bangley area.

Porphyritic trachytes

These are represented by rocks from Kae Heughs (S 4522), Peppercraig (S 616, 4519), Bangley area (S 46174–76), near Hopetoun Monument (S 620–21), Dirleton (S 10800, 51235) and Chesters (S 10802). In specimens (S 622, 50148) from Phantassie Hill, phenocrysts or microphenocrysts of pale purple augite occur and, in the matrix, generally rather altered green aegirine-augite is present as small irregular or subophitic plates. Euhedral, pale green aegerine-augite occurs as microphenocrysts in a sliced rock (S 4519) from Peppercraig. In sliced specimens of porphyritic and sparsely porphyritic trachytes from Kae Heughs (S 50146–47, 50157) the aegirine-augite occurs as trachy-ophitic plates, generally rather altered but, in two samples (S 4522, 50149) the pyroxene occurs as small irregular prisms or grains. Pseudomorphs in isotropic yellow chloritic material in a specimen (S 50148) from Phantassie resemble the pseudomorphs after fayalite in the Traprain Law phonolite. As noted by Bailey (*in* Clough and others, 1910, p. 132) some of the Kae Heughs specimens (S 635, 4522) are distinguished by the globular shape of the iron ore. Some trachytes (S 1007, 50155, 54327) contain rare phenocrysts of plagioclase indicating a trend towards trachyandesite. Rare phenocrysts of magnetite have been noted in specimens from Phantassie (S 50148, 50170) and from the Dirleton–Fenton area (S 10800, 51128–31).

Non-porphyritic trachytes

Non-porphyritic or sparsely porphyritic varieties occur at Craigy Hill (S 636, 51170), Chesters (S 51163) and Whitelaw (S 11331–33). Specimens from Kae Heughs include porphyritic and sparsely porphyritic varieties.

Quartz-trachytes

The quartz-trachytes in which quartz occurs in noticeable quantity are found mainly in the north and south. Porphyritic varieties occur in the Dirleton–Fenton area (S 1860a, 51128–31, 51234) and the Chesters–Kilduff area (S 50161–62, 51157). Some specimens (S 51128–29) from Dirleton contain plagioclase cores to potash-feldspar phenocrysts indicating a tendency towards quartz-banakite. Several of the rocks from the Chesters–Kilduff area (S 50162, 51157–58, 51162, 51165, 51172) contain pseudomorphs after pyroxene as microphenocrysts.

In the trachytic rocks, accessory minerals include rare crystals of zircon (S 11242, 51160, 51162) and frequent needles of apatite. Bailey (*in* Clough and others, 1910, p. 132) noted the presence of large clouded crystals of apatite and a large crystal of purplish apatite has been noted in one specimen (S 50155). Fluorite, colourless to pale purple in thin section, has been detected in a number of the trachytic rocks. It occurs as small intersertal patches in specimens from the Dirleton–Fenton area (S 51129), Kae Heughs (S 50157), Peppercraig (S 616) and Skid Hill (S 54625). In a rock (S 50162) from the Chesters–Kilduff area fluorite occurs in a drusy quartz vein and, in the specimen (S 50157) from Kae Heughs, it locally occurs in patches with quartz and chlorite. It has also been recorded in the trachyte sill at Eyebroughy. RWE

CHAPTER 9

Intrusions

INTRODUCTION

Intrusions of various ages cut strata of all formations in the district. The earliest are late-Caledonian minor intrusions, ranging from felsite to quartz-microdiorite, which cut Ordovician and Silurian strata. There is some evidence of a major granitic batholith almost entirely concealed under these rocks. Several periods of magmatic activity occurred during the Carboniferous, mainly intruding Devono-Carboniferous and Carboniferous strata, though the latest suite also cuts the Lower Palaeozoic rocks. Basaltic and trachytic to phonolitic sills and plugs are related to the Garleton Hills Volcanic Rocks of Dinantian age. A monchiquite-basanite suite of sills, plugs and dykes derives at least partly from a late stage of the Garleton Hills activity, but may also include intrusions of the Stephanian episode recorded in adjacent districts (Forsyth and Chisholm, 1977, pp. 179–180). Teschenite sills are related to the olivine-dolerite suite of Namurian to early Westphalian age (Forsyth and Chisholm, 1977, pp. 136–137). The W–E trending quartz-dolerite dykes are of late-Westphalian to Stephanian age (Cameron and Stephenson, 1985). A tholeiite dyke, possibly of the Tertiary suite, occurs in the district. ADM

INTRUSIONS OF LATE-CALEDONIAN AGE

Numerous thin dykes of late-Caledonian age cut the Ordovician and Silurian rocks of the district. Most of them have trends roughly parallel to the strike of the country rocks and the few which have a divergent trend are petrographically similar to those with more normal trend. None of them cut the Lower Devonian sediments in the district. The dyke-types include felsite, quartz-porphyry, acid porphyrite and porphyrite and less commonly feldspar-porphyry, plagiophyre, microgranodiorite, quartz-microdiorite and lamprophyre. Two small bosses of granodiorite occur. A few dykes of altered basic rocks occur, whose age is uncertain.

With a few notable exceptions the minor intrusions are extensively altered. Mafic minerals are almost entirely replaced and though the plagioclase is locally fresh or present as remnants, it is generally highly sericitised and apparently albitised. Secondary hematite and limonite are common in many of the rocks. As a result of the varying types of alteration, distinction of many of the types may be difficult or misleading both by macroscopic and microscopic examination. RWE

From gravity and magnetic anomalies in the vicinity of Lammer Law [524 618], Bennett (1969) suggested the presence of a roughly circular stock-like granitic intrusion, the Lammer Law 'granite', its top surface estimated at 150 to 300 m below OD. Other evidence for the buried granite comes from thermally altered dyke and sedimentary rocks in a metamorphic aureole. Lagios and Hipkin (1982) postulated that this intrusion is part of a granite batholith which underlies the whole Southern Uplands. Other possible expressions of the batholith are the Kidlaw and Stobshiel granodiorites, north-east of Lammer Law, and the Priestlaw granite some 10 km to the east in the adjacent Dunbar district. ADM

DETAILS

Felsite

The felsites are fine-grained, sparsely porphyritic or microporphyritic rocks. Quartz occurs as infrequent, euhedral or corroded microphenocrysts. Sporadic phenocrysts or microphenocrysts of feldspar occur in many of the felsites and the nature of the feldspar phenocrysts or microphenocrysts varies. In some dyke-rocks only alkali-feldspar is noted as turbid orthoclase (S 52133, 54389), or microperthite (S 48492, 48535) and rarely checker-albite is seen (S 56479, 56486). In one specimen (S 54390) phenocrysts of both potash-feldspar and sericitised sodic plagioclase occur, the latter locally mantled by potash-feldspar. In a few rocks, pseudomorphs of kaolinite probably represent original potash-feldspar microphenocrysts (S 54361–2) and locally plagioclase may be represented by pseudomorphs. The matrix of the felsites is composed, in the main, of anhedral quartz, alkali-feldspar (locally replaced by kaolinite and sericite) with small ragged flakes and sheaves of colourless or faintly greenish mica. Spherulitic structure is sparsely developed in a few specimens (S 48535, 56479, 56486) and in one fine-grained felsite (S 49222) large patches of micrographic intergrowth are developed. The matrix of one felsite (S 56453) is different in character and contains numerous, moderately flow-aligned, small, orthophyric crystals of potash-feldspar in a cryptocrystalline base. Baryte occurs as an aggregate apparently pseudomorphing an earlier phenocryst in one specimen (S 48492) and also rarely in thin veinlets of quartz. Some hematite pseudomorphs are present, probably after pyrite. A thin vein of clear albite (ca An_8) occurs in one specimen (S 56486) and apatite is an accessory mineral.

Feldspar-porphyry

These rocks (S 48494, 48529, 48544) resemble the quartz-feldspar-porphyries apart from the relative sparsity or absence of quartz as phenocrysts. They contain phenocrysts of generally intensely sericitised sodic plagioclase. Muscovite occurs as ragged plates which may, at least in part, represent altered biotite. Some pseudomorphs in micaceous aggregate apparently after hornblende occur and in this respect the rocks have affinities with those designated as acid porphyrites. The matrix is composed of anhedral lathy crystals and plates of sodic plagioclase, anhedral quartz and much fine-grained white mica. Apatite and more rarely zircon occur as small prisms.

Quartz-porphyry

The quartz-porphyries are characterised by the presence of generally numerous phenocrysts of quartz and feldspar, accompanied in most specimens by phenocrysts of colourless to faintly green micaceous aggregates interleaved with hematite or limonite and probably representing pseudomorphs after biotite. The relative proportions of the phenocrysts vary as does the type of feldspar forming the phenocrysts. In some sliced rocks (S 29306, 48493, 56472) only somewhat sericitised sodic plagioclase occurs as phenocrysts but in others potash-feldspar is also present and may predominate. The potash-feldspar is commonly rather streaky and may be sodic orthoclase or microperthite (S 56446–7, 56466, 56494). Locally all the feldspar phenocrysts are very altered and as a general rule the potash-feldspar is kaolinised and the plagioclase is sericitised. In some specimens (S 56371, 56377) scattered phenocrysts are composed of a micrographic intergrowth of quartz and alkali-feldspar. The matrix of the quartz-porphyries is composed of an aggregate or mosaic of anhedral quartz and alkali-feldspar and may be very fine-grained (S 48493) or quite coarse (S 56446–7, 56472). Locally, micropegmatite is developed in the matrix (S 56447, 56672) and two sliced rocks (S 50968–9) are extremely spherulitic.

Acid porphyrite

The acid porphyrites are characterised by numerous phenocrysts of plagioclase, commonly with biotite and in some instances with hornblende. The plagioclase is largely albite-oligoclase but it is probably albitised since rare remnants of more calcic plagioclase are seen. It is generally highly sericitised. The biotite phenocrysts are normally pseudomorphed by chlorite, carbonate and colourless mica though fresh biotite is seen in one specimen (S 56480). Hornblende is represented only as pseudomorphs varyingly in chlorite and carbonate and is more common as microphenocrysts. Corroded xenocrysts or phenocrysts of quartz occur in some specimens (S 54349). The matrix of the rocks is variable and contains small laths or plates of turbid, commonly sericitised feldspar accompanied in some by pseudomorphs after biotite flakes, or rarely fresh biotite (S 56445) and in some with small pseudomorphs after hornblende. Quartz is common in the matrix, either having granular habit (S 29307, 54338), or in a number of specimens (S 56366, 56488), occurring as locally quite large poikilitic plates. Some of the quartz may be secondary. In some specimens (S 56354, 56438), delicate micropegmatite is developed and locally, large granophyric intergrowths occur (S 52135, 56431). Rocks such as the last might be classed as porphyritic microgranodiorite. A large irregular xenocryst of garnet extensively replaced by chloritic material occurs in one sliced rock (S 56449). Apatite and more rarely zircon occur as accessory minerals.

Microgranodiorite

One dyke-rock (S 54324) has been assigned to microgranodiorite. It contains phenocrysts of oscillatory zoned andesine, partly replaced by sericite, albite and kaolin, large pseudomorphs after biotite and smaller pseudomorphs after hornblende in a matrix composed of small laths of albitised plagioclase and a mosaic of anhedral quartz and alkali-feldspar.

Porphyrite

The porphyrites are mainly extremely altered rocks. They are characterised by the presence of phenocrysts and microphenocrysts, generally of both plagioclase and hornblende, but in some, of hornblende alone. Biotite also occurs as phenocrysts in some specimens. The matrix of the porphyrites, in addition to small crystals of mafic minerals, may be composed of a plexus of laths of plagioclase (S 50973, 52707, 54379) or of small, squat tablets of plagioclase in a fine-grained base (S 52701–2) or may be extremely fine-grained (S 54345). For the most part alteration is marked; hornblende is pseudomorphed by a variety of minerals (chlorite, carbonate, micaceous aggregates, leucoxene and hematite) and the biotite is varyingly pseudomorphed by chlorite and aggregates of faintly yellowish mica with leucoxene and hematite or limonite. The plagioclase, both phenocrysts and matrix, is commonly partly or entirely replaced by sericite (S 52132, 56352) and in some instances phenocrysts are patchily kaolinised (S 49217, 56347, 56475). The kaolinisation appears to have affected the cores of the crystals. Where plagioclase is still present in the altered rocks it is normally sodic plagioclase, approximately albite-oligoclase in composition, and probably represents albitised plagioclase. Quartz occurs intersertally and locally forms poikilitic patches in fine-grained rocks (S 56373). Locally delicate intergrowths of micropegmatite are seen in the base of some sliced rocks (S 52701). In many of the specimens there appears to be a fine-grained alkaline residuum. In one sliced rock (S 49218), small chlorite pseudomorphs, possibly after orthopyroxene, occur in addition to the leucoxene-speckled chlorite pseudomorphs after hornblende.

A few notable exceptions to the generally debased condition of the porphyrite suite occur in the Lammer Law–Black Law area. One dyke (S 56476) is remarkably fresh and contains phenocrysts of fresh plagioclase, green and brown hornblende and somewhat chloritised biotite. Another dyke (S 56482) (Plate 16.2) also is very fresh though thermally altered, presumably by the underlying Lammer Law 'granite' (p.), with the development of new biotite as aggregates pseudomorphing hornblende and as flecks in the fine-grained recrystallised base. Fresh, or relatively fresh plagioclase occurs as phenocrysts (S 56475, 56491) or in the groundmass (S 52701–2) in a few other dykes from this area. The phenocrysts and microphenocrysts have marked oscillatory zoning and values of apparently maximum anorthite content ranging from labradorite (An_{52}) to andesine (An_{44}) have been determined in individual sliced rocks.

Quartz-microdiorite

A few dykes (S 48526–8) of quartz-microdiorite are composed of coarse stout laths of turbid, highly sericitised plagioclase mantled by potash-feldspar, pseudomorphs in chlorite after hornblende, intersertal quartz, some delicate micropegmatite and small plates and rods of leucoxene and limonite.

Plagiophyre

The term plagiophyre has been applied, following the definition proposed by MacGregor (1939, p. 101) to a number of extremely decomposed non-porphyritic or sparsely porphyritic rocks, possibly originally of andesitic composition. One sliced rock (S 52718) contains pseudomorphs in bastite, possibly after orthopyroxene.

Lamprophyre

The few lamprophyre dykes of the district are mainly augite-minettes (S 52703, 56360–61) containing numerous flakes of brown biotite generally with dark margins, pseudomorphs in carbonate and chlorite after prisms of pyroxene in a base of hematite-dusted alkali-feldspar, apparently both albitic and potash-feldspar, with small grains of magnetite. Feldspar ocelli are locally developed (S 56361). One sliced rock (S 52703) is very coarse in grain, contains a few xenocrysts of quartz and has some intersertal quartz. A few serpentine pseudomorphs possible after olivine occur in this rock. One highly altered hornblende-lamprophyre (S 56343) has been recorded. RWE

Lammer Law 'granite'

The metamorphic aureole of this concealed intrusion is 2 km long by 250 to 750 m wide, aligned NE–SW roughly parallel to the strike of the Ordovician greywackes and shales. In the upper reaches of the Hopes Water, there are indurated and hornfelsed greywackes and shales. At White Sled Burn [5433 6101] a dyke of porphyrite is thermally altered. Other dykes in the area do not show thermal alteration and are presumed to post-date, at least slightly, the intrusion of the granite.

Kidlaw granodiorite

This boss of biotite-granodiorite was first recorded by Simpson (1928, pp. 112–113). It is 450 m long by 150 m wide, elongated along the strike of the Ordovician rocks into which it is intruded, and lies a short distance south-east of the Southern Upland Fault at Kidlaw. The granodiorite (S 52708) is composed of coarse subhedral to euhedral laths and plates of andesine strongly zoned to oligoclase, orthoclase as plates and mantling the plagioclase, anhedral quartz and plates of biotite. The biotite is almost entirely pseudomorphed by chlorite and carbonate with hematite; the plagioclase is somewhat sericitised and locally the cores of the crystals are replaced by kaolinite. Simpson (1928, p. 113) refers to extensive albitisation but this is not a notable feature of the sliced rock. One or two poorly shaped pseudomorphs of carbonate and chlorite occur, possibly after hornblende.

Stobshiel granodiorite

A similarly elongate boss, only 200 by 50 m, occurs about 2 km to the south-west at Stobshiel. This intrusion (S 52113) is very similar in composition and alteration to the Kidlaw granodiorite though lacking the local kaolinisation of that rock. Two sliced specimens (S 52111–2) from the south-east end of the Stobshiel intrusion are highly altered, coarse-grained, acid igneous rocks composed of abundant coarse anhedral quartz, pseudomorphs in sericite after plates of feldspar, and plates of faintly yellowish mica with hematite possibly representing pseudomorphs after biotite. WT, RWE

INTRUSIONS OF CARBONIFEROUS AGE

Many intrusions cut the Carboniferous sedimentary and volcanic rocks of the district, particularly in the northern and eastern parts. Along the coast the form of many intrusions and their relationships with the country rocks are clearly displayed. Few intrusions are known from areas of Carboniferous strata in the south and west, but this may be partly due to concealment by the extensive drift cover. The thickest known intrusion, the concealed Spilmersford intrusion, was proved by the IGS Spilmersford Borehole. In inland areas the nature and form of the poorly exposed igneous rocks are unclear. Intrusive bodies may occur within areas mapped as lavas. This is particularly so with the trachytes, which show little evidence of their intrusive or extrusive nature and little lithological variation between extrusive and intrusive bodies. A few minor intrusions of Carboniferous age also cut the Lower Palaeozoic rocks.

The majority of the intrusions are sills (Plate 12). Several volcanic plugs form very prominent features of the landscape (Plate 13). Laccoliths have been identified, most notably the classic Traprain Law Laccolith (Plate 14). Dykes occur both as small irregular networks associated with vents and as part of a major dyke swarm (Plate 15).

Classification and age

The intrusions can be divided into five fairly distinct groups:

Late Westphalian to Stephanian	Quartz-dolerite dykes
?Namurian	Olivine-dolerite/teschenite sills
?Dinantian to Stephanian	Monchiquite/basanite suite
Dinantian (Garleton Hills Volcanic Rocks)	Trachytic intrusions
Dinantian (Garleton Hills Volcanic Rocks)	Basaltic intrusions

Some intrusions are clearly related in form and lithology to the Garleton Hills Volcanic Rocks, of Dinantian age. Plugs such as those at Cairndinnis and Burning Mount are feeder pipes containing the same type of rocks as the lavas. Trachyte sills as at Pencraig and Athelstaneford have similar lithologies to the trachyte lavas, while in other phonolitic intrusions, e.g. North Berwick Law Plug and Traprain Law Laccolith, lithologies not found as lavas were produced after the eruptive phase of the volcanic episode when differentiation of the alkaline magma had reached and advanced stage (Tomkeieff, 1937; MacDonald, 1975). In this latter category also are some of the monchiquite-basanite suite of minor intrusions, lithologies also not occurring as lavas. They are associated with some of the early vents along the North Berwick coast, particularly the Yellow Man Vent. Other basanitic intrusions occur in the Yellowcraig Plantation Vent which cuts later lavas of the Hailes Member, and in strata overlying the volcanic rocks along the Gullane shore (Figure 17). These may be comparable with late intrusions in Fife which are Namurian or later in age (Forsyth and Chisholm, 1977). Teschenite sills at Gosford and Gullane cut strata up to the Lower Limestone Group and a Namurian age would be reasonable (Forsyth and Chisholm, 1977, pp. 136–161). The olivine-basalt/olivine-dolerite Spilmersford intrusion is more likely to be part of this suite than directly associated with the Garleton Hills volcanic episode. The last episode of Carboniferous intrusion, the W–E quartz-dolerite dykes, are quite clearly dated as late-Westphalian to Stephanian from evidence further west and in Fife from the Midland Valley sill-complex and dyke-swarm. In the district, the dykes are known to cut strata as young as Westphalian A.

Radiometric age-determinations

Age-determinations of igneous rocks from the district have been made (Table 3). The main volcanic episode was dated as early Viséan, which agrees with the lithostratigraphy, although the basaltic lavas of the East Linton Member are demonstrably older than the Bangley Member trachyte lavas. Most of the basaltic feeder pipes are of a similar age, apart from the St Baldred's Plug which gave a questionable Stephanian age. The trachytic intrusions gave late Viséan dates immediately following the lavas, but one Rb:Sr date suggests a more immediate connection with the trachytic lavas. The date of the Dirrington Law intrusion ties in with the younger of the trachytic dates. None of the monchiquite-basanite dates make much sense whether they are regarded as late differentation products of the Garleton Hills Volcanic Rocks or as part of the olivine-dolerite/teschenite suite. Age-dates for the latter suggest an acceptable late Namurian age, and the Spilmersford intrusion could be more appropriately put in this category. The Stephanian age of the

Table 3
Radiometric age-determination of Carboniferous intrusions and volcanic rocks

Stratigraphy	Lithology/Form	Locality (Figure 17)	Minimum Age (Ma)	Source
Post-Garleton Hills Volcanic Rocks	quartz-dolerite dyke	Bangley (Ba)	288 ± 7	A
Post-Garleton Hills Volcanic Rocks	basalt/dolerite sill	Spilmersford (SP)	297 ± 8	A
Post-Garleton Hills Volcanic Rocks	teschenite sill	Ravensheugh	307 ± 5	A
Post-Garleton Hills Volcanic Rocks	teschenite sill	Gosford Bay (GB)	313 ± 9	A
Post-Garleton Hills Volcanic Rocks	olivine-basalt plug	Yellow Craig Plantation (YP)	224 ± 6	C
Post-Garleton Hills Volcanic Rocks	basanite sill	Gin Head	234 ± 7	C
Post-Garleton Hills Volcanic Rocks	basanite dyke	Gin Head	272 ± 7	C
Post-Garleton Hills Volcanic Rocks	felsite pluton	Dirrington Law	325 ± 10	C
Garleton Hills Volcanic Rocks (intrusive)	phonolitic trachyte plug	North Berwick Law (No)	323 ± 5	A
Garleton Hills Volcanic Rocks (intrusive)	phonolite laccolith	Traprain Law (Tr)	322 ± 2	A
Garleton Hills Volcanic Rocks (intrusive)	phonolite laccolith	Traprain Law (Tr)	342 ± 4	B
Garleton Hills Volcanic Rocks (intrusive)	Dunsapie basalt sill	Borthwick	326 ± 7	B
Garleton Hills Volcanic Rocks (intrusive)	basalt plug	Burning Mount (BM)	331 ± 5	A
Garleton Hills Volcanic Rocks (intrusive)	basalt plug	Yellow Craig (YC)	339 ± 9	A
Garleton Hills Volcanic Rocks (intrusive)	Dunsapie basalt plug	Cairndinnis (CA)	348 ± 8	A
Garleton Hills Volcanic Rocks (intrusive)	Craiglockhart basalt plug	St Baldred's	296 ± 7	A
Garleton Hills Volcanic Rocks (extrusive)	trachyte lava	Phantassie	342 ± 5	A
Garleton Hills Volcanic Rocks (extrusive)	trachyte lava	Skid Hill	349 ± 7	A
Garleton Hills Volcanic Rocks (extrusive)	Dunsapie basalt lava	Kippielaw	337 ± 9	A
Garleton Hills Volcanic Rocks (extrusive)	mugearite lava	Lawhead	341 ± 9	A

Sources: A – de Souza (1974), B – de Souza (1979), C – Snelling and Chan (1976, IGS internal report)

quartz-dolerite dykes is proved elsewhere. Whether the age-date quoted is for the hybrid Bangley dyke or for the continuation of the Prestonpans–Seton dyke is not clear.

DETAILS (letters in brackets refer to Figure 17)

Basaltic intrusions related to the Garleton Hills Volcanic Rocks

Burning Mount and Cairndinnis plugs (BM, CA)

Two small volcanic plugs cut sediments below the Garleton Hills Volcanic Rocks domed up by the Traprain Law Laccolith. Both intrusions are fairly well exposed, though without visible contacts. At Burning Mount there is an associated vent-agglomerate. The rock is dark grey, fine-grained olivine-basalt of Dunsapie type, with plagioclase, augite and olivine phenocrysts, and is similar in composition to, but fresher than, the numerous lavas of Dunsapie type which outcrop to the north and west. These plugs are feeder pipes for the lavas.

Saughland Sill

An intrusion [406 612] near Saughland, although far away from any others, appears related as it consists of olivine-basalt of Dunsapie type. It is poorly exposed and its relationship with the surrounding Calciferous Sandstone Measures sediments is unclear.

Plate 12 Sill of basalt, Isle of Fidra, basalt lava in foreground (D3037)

Other small, poorly-exposed olivine-basalt intrusions occur at Markle Mains (MM), possibly a plug, and at West Fenton (WF), possibly a sill. Three highly altered sills of basic rock or white trap were encountered near the base of the Calciferous Sandstone Measures in the IGS Spilmersford Borehole.

Yellow Craig Plug (YC)

East of North Berwick, the red tuffs and marls at the south-west corner of Leckenbane are cut by an oval plug of olivine-basalt. Although the contact with the tuffs is not seen, the rocks shows a pale and glassy chilled margin. The fresh rock in the centre of the intrusion is dark grey olivine-basalt of Jedburgh type in which small plagioclase and augite phenocrysts are visible. The intrusion continues to the north-east as thin basalt and agglomerate dykes, with associated small cryptovents. ADM

Petrography

Intrusive basalts thought to be contemporaneous with the Garleton Hills volcanic episode include several Dunsapie or allied types; for example, the plugs of Burning Mount and Cairndinnis, and the Saughland Sill. The Burning Mount rock locally contains very large phenocrysts of augite and the proportion of phenocrysts is variable. In some specimens (S 11842, 53420) phenocrysts of olivine, augite and plagioclase are all common but others (S 755, 820, 11841), with increased proportion of plagioclase phenocrysts, approach the feldspathic Dunsapie type of Clark (1956, p.46). The Cairndinnis Plug is composed of coarsely ophitic porphyritic olivine-dolerite (S 11937) allied to Dunsapie type, and a small anorthosite xenolith has been recorded. A finer-grained sample (S 50370) of the intrusion is a more typical basalt of Dunsapie type but variation in augite texture from prismatic to ophitic crystals occurs. The fresh basalt (S 10853) from the Saughland Sill has here been grouped with those of Dunsapie type.

Basalts of Jedburgh type form the Yellow Craig and Markle Mains plugs. The Yellow Craig basalt (S 4603, 36487) contains many stout microphenocrysts of plagioclase and a few of olivine in a fine-grained matrix. Scattered plagioclase phenocrysts occur and are locally common in one sliced rock (S 36488). In the Markle Mains Plug (S 53423) a number of pseudomorphs after olivine form large microphenocrysts and though the plagioclase microphenocrysts are not very plentiful the rock can be assigned to Jedburgh type in view of the feldspathic nature of the matrix. A coarser-grained basalt (S 53422) contains more conspicuous plagioclase microphenocrysts. RWE

Trachytic intrusions related to the Garleton Hills Volcanic Rocks

Eyebroughy ?sill (Ey)

The tidal islet of Eyebroughy consists entirely of non-porphyritic quartz-trachyte. Previous surveys described it as a lava, but it is considered more likely to be a sill. The contacts are not exposed.

Burnside and Carperstane intrusions (Bu, Ca)

Two infilled quarries at Burnside formerly exposed an intrusive ?plug of porphyritic trachyte. Barrow (*in* Clough and others, 1910, pp. 78–79) suggested that it was continuous with the sill formerly exposed at the Carperstane limestone quarry, 2 km to the NE, and possibly with the trachyte at Craigs Quarry, Dirleton, which he considered intrusive.

Bangley Dyke (Ba)

Day (1930a) first identified a dyke as the host of the large sanidine phenocrysts, up to 50 mm long, first noted in Bangley Quarry by Barrow (*in* Clough and others, 1910, p. 79). Exposures of the dyke are dependent on the current working face of the quarry. The dyke is about 3 m wide, vertical, striking approximately N–S, and with close horizontal jointing which distinguishes it from the more massively jointed quartz-banakite lava. Day described the fine-grained groundmass as basaltic and considered that basaltic magma, passing through trachytic rocks and picking up the phenocrysts of sanidine, had given rise to a hybrid trachybasalt rock.

Athelstaneford Sill (At)

Athelstaneford village lies on an outcropping ridge formed by a trachyte sill intruded into olivine-basalt lavas of Markle type. In previous surveys this was thought to be continuous with the Pencraig intrusion to the south-east, but there is no evidence for this, the area between being covered by thick drift in a buried channel and the lithologies being not very similar. Although no contacts are visible, on three sides the basalt lavas of the Hailes Member can be seen close to the intrusive, blue-grey, fine-grained, non-porphyritic trachyte.

Pencraig Sill (Pe)

The high ground from Markle Mains Heughs to Pencraig Wood is formed from the glaciated outcrop of a trachyte intrusion. Although no contacts are visible, the discordant junction can be traced against Markle basalts and mugearites of the Hailes Member to the west and south, against Dunsapie basalts of the East Linton Member to the north and against red tuffs of the North Berwick Member to the east. The contact is obscured by drift below Markle Mains Heughs to the north. The shape of the intrusion is not clear but has been variously described as a sill or laccolith. The trachyte is best exposed in the 500-m long cliff of Markle Mains Heughs, in the new Markle Quarry and in Pencraig Quarry, where it is grey-purple, fine-grained, non-porphyritic except for rare feldspar phenocrysts, and is fissile or platy near the surface. A small outcrop of trachyte at Markle Mains (Ma) may be connected with this intrusion indicating a more extensive body.

Garvald Sill (Ga)

A small part of a trachyte sill, intruded into Devono-Carboniferous strata, lies within the district near Garvald, but the main part lies within the Dunbar district to the east.

North Berwick Law Plug (No)

Undoubtedly the most prominent natural feature in the district is the conical volcanic plug of North Berwick Law (Plate 13), 187 m high and standing about 140 m above the surrounding countryside. The west face has been swept clear by ice and the east side is partly obscured by a glacial-tail feature. The country rock is basaltic tuff of the North Berwick Member, but no contacts are visible. Though rocky, there is not much good exposure on the Law, and the old quarry on the south side, from which was obtained the russet building stone that gives the old part of North Berwick its distinctive character, has almost been filled. The remaining 4-m face is in the fine-grained edge of the plug, and shows blue-grey, weathered to mottled pink-grey, fine-grained, porphyritic, phonolitic trachyte, much iron-stained and with prominent radial joints. Fresher and coarser rock occurs near the summit.

Hairy Craig Sill (Ha)

Cutting the sediments domed up by the Traprain Law Laccolith is a small sill-like intrusion with small stringers running south-west. It may be associated with the laccolith although there is no observed connection, and contacts with the sediments are not exposed. The top part of the 10-m face at Hairy Craig is amygdaloidal, suggesting near-surface intrusion, and the joints indicate a dip of 30° to the SE. The rock is a pink, fine-grained trachyte with rare feldspar phenocrysts and inclusions. ADM

Traprain Law Laccolith (Tr)

The intrusion (Plate 14) is best described as a whole although only the western half lies within the Haddington district, and the main quarry exposures occur in the adjacent Dunbar district. In plan the intrusion is pear-shaped with an axis striking at 55°. The laccolithic form of the intrusion was first described by Bailey (*in* Clough and others, 1910, p. 98). Erosion has exhumed the original form of the laccolith of hard igneous rock by stripping away the soft sedimentary cover. Intrusion of the magma domed up the Lower Carboniferous sediments and the tuffs and basalt lavas of the North Berwick and East Linton Members, as seen in the surrounding basalt trap-featuring. Evidence of sediments under the intrusion, which would prove its laccolith form, has been offered by Stillman (*in* McAdam and others *in* Upton, 1969) confirming the proposal by Bailey (*in* Clough and others, 1910, p. 98) of a flat lower surface because of the presence of vertical, columnar cooling-joints. The main evidence for the laccolith-shape consists of strong, curved, platy cooling-joints which parallel the present surface of the intrusion, the upward fining of the grain size (MacGregor *in* Mitchell and others, 1960, p. 68) and the vesicular top indicating near-surface intrusion (Upton and MacDonald *in* Craig and Duff, 1975, p. 76).

Additional light on the shape of the Traprain Law body was furnished by IGS geophysical studies in 1965. Ground magnetic profiles established across Traprain Law suggest that the contacts be tween the phonolite and the surrounding sediments, on both the north-west and south-east sides of the hill, are very steep-sided. The anomaly associated with the phonolite on the ground is very similar to an elongated, intense, aeromagnetic anomaly running south-west from Traprain Law at least as far as Whitelaw Farm [3567 6720]. This anomaly is probably produced by a buried igneous body since it bears no apparent relation to the Calciferous Sandstone Measures sediments or basalt and trachyte lavas at the surface. The relatively large amplitude and narrow width of the anomaly are indicative of a steep-sided igneous body, with a high magnetic susceptibility, situated at shallow depth. The Traprain Law phonolite has a magnetic susceptibility sufficiently high (about

Plate 13 North Berwick Law, a phonolitic trachyte plug (D2329) MNS 3965

2200 × 10^6c.g.s. units) to produce the observed anomaly and it seems probable that this anomaly is caused by an underground extension of the Traprain Law intrusion. At Whitelaw Farm the maximum possible width of this proposed extension is about 1000 m and depth determinations suggest that its top surface lies between 30 and 45 m below ground level.

The fresh, hard, well-jointed phonolite is variously pink and grey mottled or striped. Sparse large phenocrysts are oligoclase, sanidine and cryptoperthite, while small phenocrysts of clinopyroxene, altered hornblende and apatite occur. The coarse-grained matrix has a trachytic texture formed by the abundant feldspar laths of sanidine, soda-orthoclase and soda-plagioclase. Aegirine-augite, fayalite, analcime, nepheline, sodalite, apatite and iron ore make up the rest of the rock. Detailed petrographic descriptions of the phonolite have been given by Hatch (1892), Bailey (*in* Clough and others, 1910), MacGregor and Ennos (1922), Tomkeieff (1952) and Upton and MacDonald (*in* Craig and Duff, 1975).

A feature of the Traprain phonolite is the number of unusual minerals collected from veins, druses and vugs during the working life of the quarry. MacGregor (*in* Mitchell and others, 1960) lists analcime, anhydrite, apophyllite, datolite, natrolite, pectolite, prehnite, gypsum, selenite and stilpnomelane. Calcite and alkali feldspar also form good crystals. Batty and Moss (1962) recorded the occurrence of powellite.

Another feature uncovered during quarrying of the intrusion is the existence of sedimentary and igneous xenoliths. Day (1930a; 1932b) recorded a xenolith of baked shale 1.8 m across and xenoliths of sandstone 9 m and 2.7 m across, the shale and sandstone showing different thermal metamorphic effects. Small basic igneous xenoliths were described by Tomkeieff (1952) as analcime-hornblende-trachybasalt.

Colour-banding has been recorded by previous authors (Bailey *in* Clough and others, 1910; Tomkeieff, 1952), involving a paler pink rock and a darker grey rock. Their relationship as flow-banding is spectacularly displayed in the east face of the now-abandoned quarry. The contact between dark and light rock is generally sharp. The bands retain separate identity over the 10-m high face of the quarry where convection currents in the magma have swirled the bands into complex patterns. Preservation of the delicate convection structures suggests proximity to a feeder pipe at the north-east end of the laccolith. It is clear that the colour-banding has no relationship with the cooling joints, as the two features intersect at varying angles.

Bailey regarded the banding as due to alteration of the fresh transparent sanidine laths in the dark rock to decomposed feldspar in the pale bands. Tomkeieff agreed with Bailey, and Upton and MacDonald (*in* Craig and Duff, 1975) suggested a slight hydrothermal effect to account for the alteration. At this time, 1984, there is a danger of the classic scientific and educational locality of Traprain Quarry, already a Site of Special Scientific Interest, being filled in or 'landscaped'. It is hoped that efforts of local bodies will preserve the quarry as a nature reserve. ADM, RWE

Petrography

The trachyte intrusions include varieties represented among the lavas. The Burnside intrusion is composed of porphyritic trachyte (S 53429) containing phenocrysts of clear potash-feldspar, some with cores of plagioclase. The groundmass feldspar occurs as coarse elongate slender laths, about 0.5 mm long, which are markedly flow-aligned. Scattered chlorite pseudomorphs after euhedral pyroxene occur as microphenocrysts. The intrusions of

Plate 14 Traprain Law, a phonolite laccolith (D1850)

Eyebroughy, Athelstaneford, Pencraig and Garvald are non-porphyritic varieties in which quartz occurs as granular patches or locally (S 11834) as large patches against which the feldspar is euhedral. The textures vary from intrusion to intrusion. The Pencraig trachyte is generally characterised by squat, somewhat orthophyric tablets of alkali-feldspar generally with poor to moderate flow-structure (S 628, 10854, 52087–88), and pseudomorphs after subophitic pyroxene occur. In a fine-grained rock (S 53413) from this intrusion the feldspar occurs as small laths with better flow alignment. Sliced rocks (S 49347, 50347) from the Garvald Sill are characterised by slender small flow-aligned laths of alkali-feldspar. In the trachyte of the Eyebroughy Sill the feldspars varyingly occur as coarser slender rectangular laths with marked flow-orientation and the development of lines of 'ruck' (S 10827, 33554), or as rather squat, orthophyric tablets (S 33553) as in the Pencraig Sill. Apatite occurs as small colourless needles and as larger crystals but a few stout crystals of dusky apatite, some with striae of inclusions, have been observed in specimens from the Garvald (S 49347, 50347), Burnside (S 53429) and Pencraig (S 53413) intrusions. Fluorite, recorded in some of the trachyte lavas, has been noted as rare anhedral patches in a sliced rock (S 33553) from the Eyebroughy intrusion.

The trachybasalt dyke cutting the lavas of Bangley Quarry contains large phenocrysts or xenocrysts of glassy sanidine with a few phenocrysts of decomposed plagioclase in a matrix of decomposed feldspathic olivine-basalt (S 13246, 36689, 46172, 51190). Though classed as a trachybasalt the rock may be regarded as a xenocrystic decomposed olivine-basalt.

In the phonolitic trachyte of North Berwick Law alkali feldspar occurs locally as generally altered phenocrysts (S 639, 51239) but more commonly as stout tablets forming microphenocrysts like those of the Bass Rock (S 834a, 835, 10797). The potash-feldspar of the matrix generally occurs as elongate, slender, flow-aligned laths (S 4597, 51237–39). Lines of ruck are commonly developed. Microphyric varieties contain large laths of feldspar in a matrix of small microlites (S 4521). In two specimens (S 834a, 10797) anhedral plates and prisms of aegirine-augite occur and pseudomorphs probably after fayalite. Generally the pyroxene is completely altered. Analcime occurs in intersertal patches but some patches (S 834a, 4521, 51239) may represent pseudomorphs after nepheline. Purple fluorite has been noted in several thin sections (S 834a, 4521, 51239) (Plate 17.1) in small patches and locally with kaolin in part-altered feldspar microphenocrysts.

The Traprain Law phonolite has been described in detail by Bailey (*in* Clough and others, 1910, pp. 128–129), MacGregor and Ennos (1922) and Upton (*in* Sutherland, 1982, p. 517). The rock is commonly aphyric but locally contains scattered phenocrysts or microphenocrysts of alkali-feldspar (sanidine-microperthite) and oligoclase, the latter mantled by alkali-feldspar (S 35182, 63533–34). Locally small phenocrysts of brown amphibole occur (S 34230–31) with a part-resorbed rim of granular ore and clinopyroxene but more commonly (S 4523, 35181) they are pseudomorphed by ore. The matrix is composed of laths of potash-feldspar, aegirine-augite, fayalite (or pseudomorphs) and feldspathoids (nepheline, analcime and sodalite). The aegirine-augite generally occurs as anhedral, locally subophitic plates which may be as large as 1.4 mm diameter (S 4525, 21173) and in some varieties as smaller subhedral prisms (S 634, 997, 4527–28, 35181). The fayalite is commonly fresh and mainly forms trachy-ophitic plates exceptionally as large as 1.5 mm diameter (S 21173), but also occurs as smaller anhedral plates (S 21805, 63525) and rarely as small subhedral to euhedral crystals (S 633, 997, 4529). Nepheline occurs as poorly shaped crystals and analcime and sodalite occur intersertally. The isotropic feldspathoids are much more prominent as clear intersertal patches in the more altered pink phonolite and the sodalite commonly margins such areas. Apatite occurs as clear needles and crystals and also dusky crystals. The intrusion has long been known as a locality for minerals in vugs and veins (p. 63) and many of these also occur on a small scale in thin sections. Xenoliths of dark fine-grained igneous rocks have been recorded in the quarry and feldspathic varieties have been described by Bailey (*in* Clough and others, 1910, p. 129) and Bennett (1945, p. 41), and an analcime-rich variety (analcime-hornblende-trachybasalt) has been described in detail by Tomkeieff (1952). Bennett noted the similarity of the inclusions to the hornblende-kulaite lava and Tomkeieff (p. 369) suggested that they may derive from a pre-existing plug of such material.

The rock of the Hairy Craig Sill is a fine-grained trachyte with scattered small phenocrysts of alkali-feldspar and rarely of oligoclase (S 11355). Some euhedral microphenocrysts of brown hornblende occur which are partly or completely resorbed and replaced by iron ore and a few partly resorbed crystals of biotite with a rim of iron ore occur (S 11356, 53414). Analcime, clear and turbid, occurs interstitially and in small patches. Bailey (*in* Clough and others, 1910, p. 129) noted (S 11355) that certain zeolitic patches occur with concentric zonal arrangement of feldspar microlites. RWE

Basanite-Monchiquite suite

North Berwick Abbey Plug (NB)

A poorly exposed, oval intrusion of monchiquite which forms The Knoll at North Berwick is taken to be a plug intruded into basaltic tuffs of the North Berwick Member.

Yellow Man dykes (ZV)

In the vicinity of and within the Yellow Man Vent are several irregular basanite dykes and a small plug, which represent feeder channels for the vent. The dykes vary greatly both in width, from 1 to 6 m, and in strike, with right-angle bends and splitting into offshoots. In the intertidal zone the dykes form ridges standing proud from the soft tuffs and coarse agglomerates of the vent (Plate 15) though in some cases the indurated tuff forms the feature. Prominent cooling joints within the dykes lie parallel to their margins. The rock is a black, fine-grained, non-porphyritic basanite.

The Leithies Sill (LE)

This intrusion was described in great detail by Day (1925a) who interpreted it as a laccolith. The country rocks are green, bedded tuffs, tuffaceous sediments and cementstones of the North Berwick Member. The intrusion is exposed in the intertidal zone and erosion has dissected it into several isolated parts. The basal contact is

Plate 15 Basanite dyke cutting vent-agglomerate, Yellow Man Vent, North Berwick (D1113)

well exposed and a few outliers of sediments which form the roof are preserved. A few metres thick in the south, the intrusion is over 10 m thick in the north. Along contacts with sediments, the intrusion shows vesicles, alteration to white trap, and inclusions of sedimentary material. Where fresh the rock is a well-jointed, black, fine-grained basanite.

Bonnington and New Mains intrusions (BO, NM)

These are two poorly exposed little-known intrusions, possibly sills. Both were once quarried and the outcrops and fragments available show them to consist of black, fine-grained, jointed basanite.

Chesters Quarry and Limplum sills (CQ, LI)

These two sills, among the largest intrusions in the district, lie between the Dunbar–Gifford Fault and the Southern Upland Fault. Although close together, they cut into Upper Old Red Sandstone sedimentary rocks and basaltic tuffs of the North Berwick Member. The sills are generally poorly exposed, their contacts not usually visible, and their thickness uncertain. The rocks consist of black, fine-grained basanite.

Kidlaw ?plug

This basanite intrusion [506 647] was formerly considered a sill (Bailey *in* Clough and others, 1910, pp. 100, 106). Simpson (1928, p. 111) suggested that the basanite was a plug in a vent, a view supported by more recent geophysical evidence (Bennett, 1969). Kidlaw Quarry [507 643] has a 7-m section of basanite, but elsewhere exposures are poor and the contacts not seen. The dark, fine-grained basanite is notable for numerous inclusions of lherzolite (olivine, enstatite, diopside, picotite) which weather out as holes, and for speckling due to poikilitic patches of analcime 1 mm or more across (McAdam and others *in* Upton, 1969, pp. 42–43). Bennett (1945), who also thought the intrusion a plug, suggested the rock contained leucite pseudomorphs.

Ewingston ?sill

Magnetic traverses across the basanite intrusion [494 643] forming the hilltop near Ewingston, 1.5 km W of the Kidlaw plug, indicated that this body may be a sill (Bennett, 1969).

Intrusions in the Silurian

Two monchiquite intrusions cutting Silurian greywackes and shales have been recorded in the south-west corner of the district. Both have large xenocrysts of biotite and xenoliths of shale and greywacke. The more northerly, in the Soonhope Burn [534 576], has been mapped as an elliptical body 80 m long and 40 m wide. The second, near the head of Thorny Cleugh [546 551], a tributary of the Whalplaw Burn, appears to be a dyke.

Black Rocks and Cheese Bay sills (BR, CB)

At these localities the outcrops of olivine-basalt of Hillhouse type are sufficiently similar to suggest they may be part of the same intrusive body (Figure 10). The complex nature of the deeply-eroded Cheese Bay Sill was described in detail by Day (1916a), but little has been published on the Black Rocks Sill. The country rocks are dolomitic sediments, in places tuffaceous, belonging to the Calciferous Sandstone Measures above the Garleton Hills Volcanic Rocks. On the shore the lower contact of the Cheese Bay Sill is well exposed, and a few outliers of the roof are preserved. Contact phenomena with the dolomitic sediments include vesicles and sedimentary inclusions within the sill, alteration of the sill to white trap, and brecciation and folding of the sediments. Columnar cooling structures are well developed in the sill opposite Eyebroughy.

Dirleton ?plugs (DI)

Inland of the last localities, several knobs of the basaltic rock stick through the blown sand cover. These appear plug-like, but being similar in lithology to the sills on the shore, they may be part of the same sill-complex.

Longskelly Sill (LO)

This sill-like body is intruded into the agglomerate of the Yellow Craigs Plantation Vent. Irregular contacts of the sill with the agglomerate are visible on its west and south margins, but the other contacts are obscured by sand (Day, 1932c). The sill consists of olivine-basalt of Hillhouse type, very similar in composition to the nearby sills just described.

Fidra and Lamb sills (FI, LA)

Both islands are entirely composed of intrusive olivine-basalt, and taken as part of the same flat-lying sill, though no contacts or country rocks are visible (Plate 12). Vertical columnar joints cut by near-horizontal joint planes are common on Fidra. On the Lamb the columns are smaller, about 30 to 50 cm diameter, are even better developed, and dip to the ESE, showing a change of dip from 80° in the west to 45° in the east. The rock is variable, being generally an olivine-basalt of Dalmeny type, tending in places to basanite. On Fidra, nodules of peridotite and pyroxenite are a feature of the rock. Xenoliths of sedimentary rocks also occur.

Yellow Craig Plantation Plug (YP)

The sharp, 10-m high hill of Yellow Craig is a plug within the Yellow Craig Plantation Vent, although its relationship with the vent agglomerate is not seen. The rock is a dark grey, fine-grained, non-porphyritic olivine-basalt of Dalmeny type, similar to the rock of the nearby sills of Fidra and the Lamb. ADM

Plate 16 Photomicrographs of thin sections of igneous rocks. All plane polarised light except 5 which is crossed polars.

1 Phonolitic trachyte (S 51239), intrusion, North Berwick Law quarry. Large rectangular phenocryst of alkali-feldspar (centre) part replaced by fluorite (high relief) × 27
2 Thermally altered porphyrite (S 56482), dyke, White Sled Burn. Phenocrysts of biotite and plagioclase, small euhedral hornblende phenocrysts replaced by biotite aggregate × 25
3 Basanite allied to Hillhouse type (S 11843A), sill, Sheriffside quarry. Augite 'spherulite' with intersertal glass × 60
4 Basanite (S 52719), dyke, Kidlaw Burn. Analcime pseudomorph (centre) after well-shaped crystal of leucite with concentric inclusions × 150
5 Ocellar camptonitic basanite (S 56345), dyke, near Fairnylees. Composite xenolith of kaersutite and enstatite (at right), part of large magnetite xenocryst (top left) × 25
6 Trachyte (S 50146), lava, Yellow Craigs. Pronounced trachytic flow-structure of small feldspar microlites interrupted by line of ruck (left side) × 23

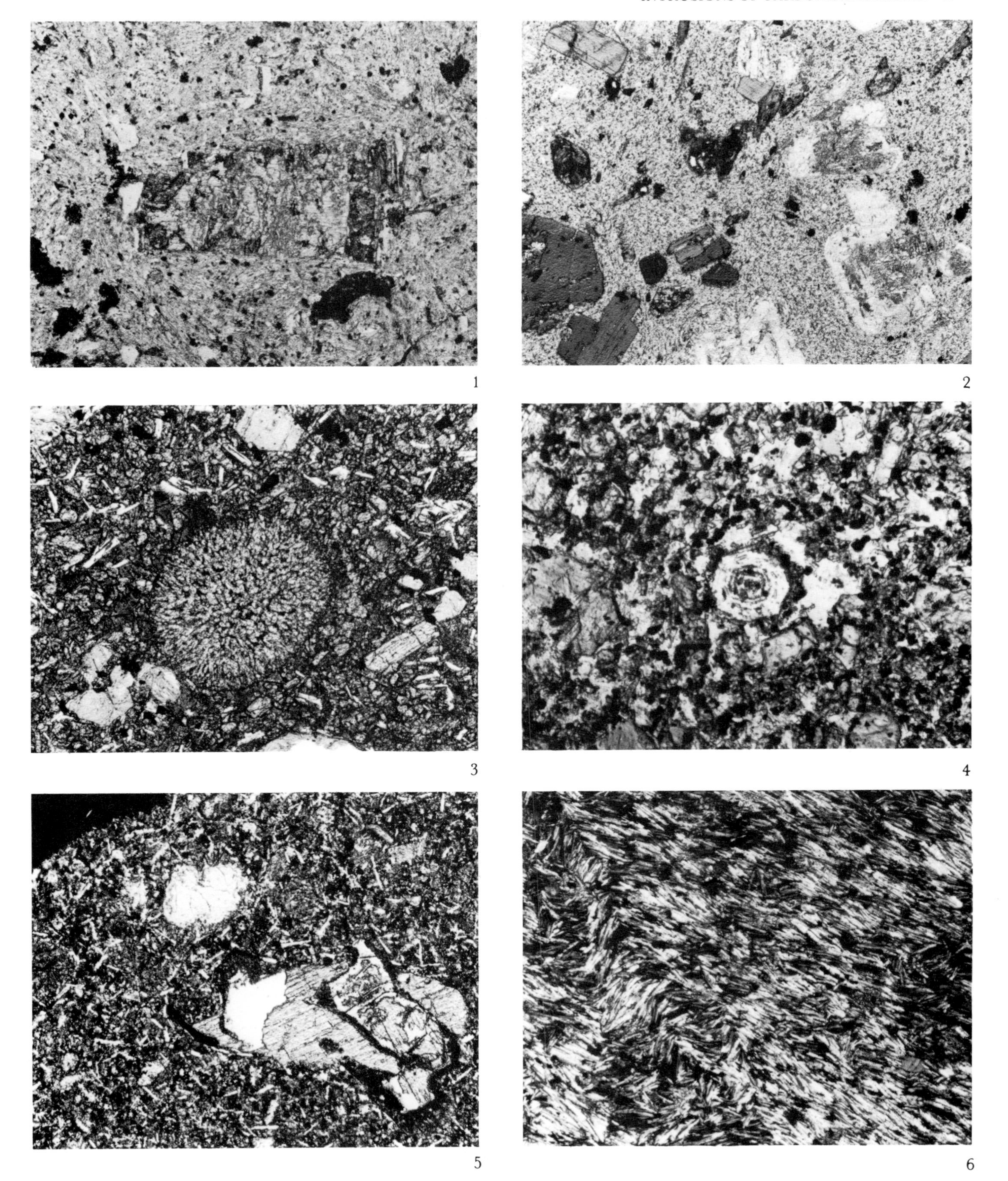
1
2
3
4
5
6

Petrography

The basanite suite ranges from rocks allied to basalts of Hillhouse type to nepheline-basanites and analcime-basanites and includes rocks described as monchiquites by Bailey (*in* Clough and others, 1910, pp. 106–107). In current usage these rocks, in view of the absence of hornblende, are not monchiquites (Elliot *in* Forsyth and Chisholm, 1977, p. 207–210) and terms such as pyroxenic analcime-basalt or olivine-analcimite are more appropriate for such feldspathoidal rocks. Also grouped with this suite of intrusions are some less pyroxenic rocks more closely allied to Dalmeny type such as the basalt sills of Fidra and the Lamb.

Two intrusions of chilled 'monchiquitic' rocks are apparently free from feldspar but it may be that they represent chilled basanites rather than chilled olivine-analcimites.

Nepheline-basanite forms the Chesters Quarry Sill and part of the Limplum Sill. The Chesters Quarry rock (S 1005, 54624) contains numerous phenocrysts of olivine and purple augite in a matrix with many prisms of purple augite and grains of magnetite. Bailey (*in* Clough and others, 1910, p. 110) noted the presence of orthoclase with a variety of nepheline which he described as 'nepheline-x' the latter euhedral against areas of analcime. Stout prisms of apatite occur and scattered flakes of biotite are present. One sample (S 49349) is finer in grain and contains irregular leucocratic segregation-veins composed of stout plates of plagioclase, pseudomorphs probably after nepheline, a few prisms of augite and plates of magnetite in a limpid faintly brown glassy base with augite crystallites. Bailey (p. 106) noted that the Limplum sill contains an increased proportion of feldspar as traced to the south. Specimens from Limplum, Townhead, Baro Wood and Sheriffside are all characterised by phenocrysts of olivine and less commonly of purple augite in a matrix rich in small prisms of purple augite and grains of magnetite. In specimens from Limplum (S 11351, 49352) clear analcime forms large areas in the base with poikilitic plates of orthoclase and, in one rock (S 49352), with large poikilitic plates of rather altered nepheline. A little anhedral oligioclase occurs in one specimen (S 11351). Quite large irregular flakes of biotite occur and also (S 49352) some blades of hornblende. In specimens from Townhead (S 48518), Baro Wood (S 11352) and Sheriffside (S 11843) plagioclase occurs as numerous tiny laths in a base of clear analcime (S 11352) or of pale brown glass with analcime in patches (S 11843, 48518) and these rocks form a link with basalts of Hillhouse type. In sliced rocks from Sheriffside (S 48516) and Yester (S 10744, 48476) the augite phenocrysts are pale in colour as is the augite of the groundmass and the rocks are more leucocratic and have more numerous microlites of feldspar.

Analcime-basanites from The Knoll (S 10826, 36483) and New Mains (S 51134) intrusions both contain many small phenocrysts of olivine, or its pseudomorphs, and rarely of purplish brown augite. The Knoll basanite contains much analcime and little plagioclase but that from New Mains contains more numerous plagioclase microlites. In the basanite of The Leithies (S 36771), a rather altered rock, microlitic augite is plentiful in the matrix. Feldspar laths are rare and Day (1925a, p. 301) describing a presumably fresher sample, regarded the rock as a monchiquite. In the basanite of the Bonnington Sill (S 54097) the groundmass contains, in addition to microlitic augite and granular iron ore, small laths of plagioclase with a little alkali-feldspar, some zeolitic patches possibly after nepheline and areas of analcime. The analcime is locally concentrated in large areas to the almost complete exclusion of feldspar (cf. Kidlaw basanite, below). The Yellow Man basanite (S 1651) contains in addition to the normal phenocrysts of olivine and augite a number of lathy microphenocrysts of plagioclase and has some resemblance to basalt of Dunsapie type.

The distinctive basanite of Kidlaw type was described in detail by Bailey (*in* Clough and others, 1910, pp. 111–112). The rock is typically less porphyritic than the other basanites though the locality is well known as a source of spinel-lherzolite xenoliths, and occasional large phenocrysts or xenocrysts of pyroxene occur (S 11949). Microlitic augite is predominant in the groundmass with feldspar, biotite, analcime and magnetite octahedra. As noted by Bailey the analcime forms large patches or crystals (up to c. 4 mm diameter, in thin section) poikilitically enclosing earlier minerals. Elsewhere the rock contains feldspar, mainly orthoclase but with some oligoclase. Some pseudomorphs after nepheline may occur.

Several small intrusions of basanite and feldspar-free monchiquite (or olivine-analcimite) cut the Lower Palaeozoic rocks. One intrusion (S 52719) (Plate 16.4) in Kidlaw Burn [5173 6315] contains small analcime areas, with concentric inclusions, possibly representing leucite, and has a xenocryst of bright green spinel unlike the normal brown spinel of the lherzolites. Xenocrysts of biotite have been noted in an intrusion (S 55640) in Soonhope Burn [5347 5767] and in another (S 55641) to the south in a tributary of Thorny Cleugh [5463 5507]. Kaersutite occurs as large xenocrysts in a dyke (S 56345) (Plate 16.5) of ocellar camptonitic basanite near Fairnylees [4985 5639] and a composite kaersutite-enstatite xenolith is present. In this rock kaersutite needles occur in the ocellar areas.

The Black Rocks, Cheese Bay and Longskelly sills are basalt of Hillhouse type. Varieties with a brown glassy mesostasis (S 753, 1000, 11903, 33566) resemble the glassy olivine-basalts of Hillhouse type recorded as vent intrusions in Fife (Elliot *in* Forsyth and Chisholm, 1977, p. 211). Some specimens (S 11903, 33556) from the Cheese Bay Sill contain sufficient analcime to justify identifying the rock as basanite.

Other intrusions allied to basalts of Dalmeny type form the plug of Yellow Craig Plantation, and the sills of Fidra and the Lamb. The Yellow Craig Plantation rock (S 33575, 57001) contains a few olivine phenocrysts, is locally analcimised (S 33576) and a small fragment of spinel-lherzolite (S 57001) has been recorded. Some specimens from Fidra (S 51907) contain much analcime in the matrix, but not to the exclusion of feldspar. Some of the larger crystals of pyroxene may represent disaggregated xenoliths of pyroxenite which are locally common.

Bailey (*in* Clough and others, 1910, p. 109) noted the presence, in many of the basanites and related rocks, of augite in radiating clusters, the centres filled varyingly with analcime, feldspar or glass (S 11843A, 11903) (Plate 16.3). Though somewhat similar structures in basanites in east Fife form reaction rims of quartz xenoliths or xenocrysts (Elliot *in* Forsyth and Chisholm, 1977, p. 211) such xenolithic cores have not been detected in the East Lothian examples.

Xenoliths of ultrabasic rock have been recorded from a number of intrusions. They were noted by Bailey (*in* Clough and others, 1910, pp. 107–109) as 'olivine-nodules' and were observed to occur plentifully in the Kidlaw basanite intrusion (spinel-lherzolite) and also in the sills of Fidra and the Lamb. Chapman (1976) recorded spinel-lherzolite and wehrlite nodules in the Fidra intrusion and noted the presence of sanidine megacrysts. Upton and others (1983) further recorded composite wehrlite-lherzolite xenoliths, and xenoliths of clinopyroxenite, anorthoclasite and magnetite-apatite-rock from Fidra. Ultrabasic inclusions including altered biotite-rich ultramafic rocks (glimmerite and mica-peridotite) have been recorded from the Partan Craig Vent by Graham and Upton (1978, p. 219). In the present survey a composite kaersutite-enstatite xenolith was found in one of the dykes within the Lower Palaeozoic area and some large xenocrysts of biotite have also been recorded in dykes in that area. Upton and others (1983) suggested that the variation in ultrabasic xenoliths reflect variations in the composition of the upper mantle or of cumulates at that level and that the presence of potassic hydrous minerals suggests potash metasomatism within the mantle. Other xenocrysts or xenoliths such as anorthoclasite probably formed in higher-level subcrustal or crustal magma chambers.

RWE

Olivine-dolerite and teschenite suite

Spilmersford intrusion (SP)

This intrusion is known from the 114.5-m thick section proved in the IGS Spilmersford Borehole (McAdam, 1974). Geophysical evidence indicates a laccolith body, aligned NNE to SSW. Lack of internal contacts and the small and gradual variations in composition suggest that there was a single intrusion of magma. A platy structure is developed with an average dip comparable to the regional dip of 20°. Phenocrysts, which are generally scarce, are mainly augite and olivine, with rare plagioclase, in a basaltic groundmass. Amygdales, which are numerous at the top and base of the intrusion, are commonly filled with carbonate, less often with chlorite, analcime and quartz. In overall composition the rock is an olivine-basalt which, where porphyritic, approaches Craiglockhart type.

Gosford Bay Sill (GB)

This teschenite sill cuts the stratigraphically highest strata of any intrusion in the district. It is intruded into the gently synclinal Lower Limestone Group sediments of Gosford and outcrops on the shore at the north and south ends of Gosford Bay. The lower contact with the sediments is well exposed at both localities, but the upper contact is generally obscured. Cryptovents are developed in the sediments immediately overlying the sill. The rock is a dark blue-grey, fine- to medium-grained teschenite, which is well jointed and in places roughly columnar.

Upper and Lower Gullane Head sills (UG, LG)

West of Gullane two westerly-dipping sills are intruded into the Carboniferous sediments above the Garleton Hills Volcanic Rocks (Figure 10). Contacts with the sediments are well exposed on the shore, but obscured inland. At Hummel Rocks in the lower sill, complex intrusion-structures present include calcite veining, dykes of tuff and breccia, and vesicular white trap (Clough and others, 1910, pp. 101 – 102; Day, 1914; Francis, 1960). The rock is a dark grey, medium- to coarse-grained teschenite, commonly showing spheroidal weathering.

Point Garry Sill (PG)

A small teschenite sill on the shore west of North Berwick is in truded into the basaltic tuffs of the North Berwick Member near the junctions of red tuffs, cementstone bands and green tuffs. The fresh rock is dark grey, fine-grained teschenite, much of which has a pale green to white colour due to pneumatolitic alteration partly or wholly to a white trap, with kaolin and chlorite (Day, 1932a).

Craigleith Laccolith (CR)

The island of Craigleith is entirely formed of a laccolith of trachybasalt or fine-grained essexite. The relationship with the surrounding rocks is not seen. Irregular flexures, indicated by prominent joints well developed all over the island, suggest a laccolithic form and that the present shape of the island is probably similar to that of the original intrusion. The rock is grey with pink segregations, fine-grained, holocrystalline and non-porphyritic. ADM

Petrography

The Gosford Bay Sill is composed of a very fresh olivine-analcime-dolerite (S 31852 – 53, 31856, 31862) with plentiful crystals of largely fresh olivine some of which are noticeably elongate (S 11859), laths of zoned labradorite, small subhedral to euhedral prisms of purplish augite, small plates and rods of iron ore and intersertal analcime. The analcime may be clear or turbid and is generally much smaller in amount than in the Lower Gullane Head Sill. Locally the feldspar is analcimised. In grain-size the rocks are on the border of dolerite and basalt. In one coarse specimen (S 63443) the augite occurs as plentiful stout subophitic plates.

The Lower Gullane Head Sill (S 11839, 46909 – 12, 48482), described originally by Young (1903) and Bailey (*in* Clough and others, 1910, p. 114 – 115) is a coarse teschenite containing pseudomorphs in serpentine and bowlingite after olivine, coarse laths (up to 0.9 mm long) of locally analcimised, zoned labradorite, coarse subhedral, subophitic to ophitic plates of pale purple augite and plates of iron ore. Analcime occurs in intersertal patches, in places replaced by chlorite. Potash-feldspar mantles the plagioclase adjacent to analcime areas.

The Upper Gallane Head Sill (S 11901) was described by Bailey (*in* Clough and others, 1910, p. 115) as a weathered teschenite in which the interspaces are entirely filled by chlorite. A fresher sliced rock (S 46926) from the intrusion is a fresh olivine-dolerite with strongly ophitic purple augite and neither patches of analcime nor chlorite are present.

The Point Garry Sill (S 1001, 10825, 36477 – 78) is a fine-grained doleritic rock with infrequent small pseudomorphs after olivine, ophitic plates of pale purple augite, laths of turbid plagioclase, plates of iron ore and intersertal patches of analcime and chlorite. Potash-feldspar is locally plentiful mantling the plagioclase and in leucocratic areas.

The rock of the Craigleith intrusion has been described in considerable detail by Bailey (*in* Clough and others, 1910, pp. 116 – 118) who called it a fine-grained essexite. The rock (S 11897) contains a number of small phenocrysts of pale augite (some corroded) and olivine in a matrix of numerous euhedral prisms and plates of pale titaniferous augite, scattered, small, generally serpentinised olivines, crystals of magnetite and laths, up to about 0.3 mm long, of plagioclase mantled by orthoclase. Flakes of biotite occur. Bailey noted similarities to the Kidlaw basanite in the presence of areas of analcime in the base. Several areas of zeolite occur, some of which appear to represent pseudomorphs after nepheline. Bailey noted that the Craigleith rock is coarser than the basanites, such as that of Kidlaw, and that it is a typical essexite apart from its rather too fine-grained texture. In some respects, including the presence of two types of feldspars, it is akin to the nepheline-basanite of Chesters Quarry. Analyses of the rocks have been published by Clough and other (1910, p. 117) and Day (1930b, p. 265). RWE

Quartz-dolerite suite

Quartz-dolerite dykes in the district belong to the swarm of W – E dykes which cross the Midland Valley and fed the Midland Valley sill-complex. These have generally been recognised to be of late Carboniferous or early Permian age (Macgregor and MacGregor, 1948, pp. 74 – 75). More recently these have been age-dated as 290 to 295 Ma, figures which indicate an early Stephanian or possibly late Westphalian age (Cameron and Stephenson, 1985). At Port Seton the dykes cut strata up to the Lower Coal Measures (Westphalian A) and are themselves cut and sinistrally displaced by the Crossgatehall Fault (Figure 16).

At least three quartz-dolerite dykes of the swarm cross the centre of the district. Their strike is about E10°N and their width is 40 to 50 m. The dykes are best known in the west around Port Seton and Longniddry, but isolated outcrops of the same dykes occur in the Garleton Hills and they are again seen further east at Dunbar. On the south edge of the district are isolated outcrops of another dyke with the same trend cutting Silurian rocks.

As well as onshore outcrops, the dykes were encountered offshore in undersea workings from the Prestonlinks and Prestongrange collieries. Intrusion of the dykes has burnt the coal seams and the igneous rock has been bleached and altered to 'white trap' where it was in contact with the coal. Organic gases driven out of the coal by the heat of the intrusion converted the already crystalline dolerite into this yellow to nearly white rock composed mainly of carbonates of lime, magnesia and iron, with kaolin.

Offshore dyke (PL)

The most northerly dyke was proved some 600 m offshore of Port Seton in undersea workings. The same dyke may be exposed on the shore at Gosford Bay.

Port Seton – Spittal Dyke (PS)

This dyke was also encountered in undersea workings and crops out in, and to the west of, Port Seton Harbour. Displaced north by the Crossgatehall Fault, the dyke forms the reefs of Long Craig, east-north-east of Port Seton Harbour, and Dalskelly Craigs, north-west of Longniddry. Inland it continues eastwards as far as Spittal, where it was quarried. An outcrop of possibly the same dyke occurs 4 km further east between Chesters and Kilduff.

Prestonpans – Seton Dyke (PR)

Lying 1.5 km to the south of the last, this dyke is also displaced by the Crossgatehall Fault. The intrusion has been traced east to Solomon's Tower, 1 km SE of Longniddry, and 100 m S of this locality is an outcrop of an offshoot or displaced part of the dyke. Further east, three sections of this dyke are exposed east of Bangly Hill, east of Skid Hill and near Athelstanesford (Figure 17). Surprisingly the dyke is not seen in the well-exposed ground between these first two localities. ADM

Kelphope Dyke

About 20 km further south, several outcrops are evidence of another quartz-dolerite dyke, cutting Silurian greywackes and shales, trending at E10°N and on a line with a dyke both to the west, just south of the district boundary, and to the east in the Dunbar district. In the Kelphope Burn [5148 5579] the decomposed spheroidal-weathering dyke, 30 m wide, is intruded into Silurian greywackes and shales. Another outcrop of the dyke occurs in a tributary stream 550 m to E10°N. No exposure of the dyke has been seen in the Soonhope Burn further east but in the next valley, in the Whalplaw Burn there is an outcrop of a quartz-dolerite dyke locally trending N – S.

Some other basic dykes, cutting Ordovician greywackes and shales have a tholeiitic appearance and may be altered members of the same suite.

Petrography

The offshore dyke is a carbonated fine-grained tholeiitic quartz-dolerite (S 35998) with laths of fresh, zoned labradorite, completely carbonated, subophitic pyroxene and skeletal plates of iron ore. There is a devitrified glassy mesostasis containing crystallites of plagioclase in a cryptocrystalline quartzo-feldspathic base with patches of quartz and locally this material fills amygdales. A sliced rock (S 31854) from the probable continuation on land, at Gosford Bay, is fresher, contains residual fresh clinopyroxene, and quartz is more common.

The Port Seton – Spittal Dyke is a coarse tholeiitic quartz-dolerite (S 31857) with subhedral to euhedral plates of pale clinopyroxene. There is a mesostatis of altered devitrified glass locally with cryptopegmatite and some patches of quartz. In an altered specimen (S 812) from Longniddry the pyroxene is completely carbonated and chloritised. A few serpentine pseudomorphs after euhedral olivine occur in a sliced rock (S 50168) from Kilduff Hill.

The Prestonpans – Seton Dyke also is for the main part a coarse tholeiitic quartz-dolerite (S 46181) with pale biscuit-coloured clinopyroxene and having an intersertal glassy mesostasis with patches of quartz and some cryptopegmatite. One specimen (S 50151) is more holocrystalline and contains delicately intergrown micropegmatite with patches of quartz. Pseudomorphs in bowlingite after poorly shaped ferromagnesian minerals occur in several specimens (S 31858 – 59, 50251). Locally the mesostasis glass fills steam holes (S 50153). In more chilled rocks (S 1689, 46173, 50152) the iron ore occurs as rods and as crystallites forming a delicate reticulate mesh. Pseudomorphs after euhedral olivine occur in some of the chilled rocks.

Macdonald and others (1981, pp. 59 – 65) in a paper on the geochemistry of the late Palaeozoic quartz-dolerite dykes of Scotland, gave analytical data for the East Lothian dykes. They noted that the Port Seton (or Longniddry) Dyke exhibits internal differentation from a mafic chill-zone to a centre approaching tholeiitic andesite in composition, and that, with the exception of Al_2O_3, the chemical variations are identical to those of the whole body of their data for the central Scotland dyke-suite.

The Kelphope Dyke (S 56365, 57054) is altered and contains stout laths and plates of somewhat kaolinised, albitised plagioclase, pseudomorphs in chlorite and carbonate after subhedral plates and prisms of pyroxene, and plates and rods of iron ore partly altered to hematite and leucoxene. Alkali-feldspar mantles the plagioclase adjacent to intersertal patches of quartz. Local patches of fine-grained quartzo-feldspathic residuum occur. A few amygdaloidal patches of quartz and carbonate are present.

Those dykes (S 65468, 57055) symbolised as K^D on the geological map and cutting the Ordovician are basaltic rocks which though highly altered have a somewhat tholeiitic appearance and may represent altered, fine-grained members of the quartz-dolerite suite. It is also possible that some of the extremely altered rocks (S 52705, 54391) in the same area, which have been symbolised as K on the map, may represent rotten members of the same suite.

INTRUSION OF ?TERTIARY AGE

A dyke (S 56440) of tholeiitic olivine-basalt which crops out in Soonhope Burn [5355 5929] is composed of subophitic to ophitic plates of pale biscuit-coloured augite, laths of zoned labradorite, small bowlingite pseudomorphs probably after poorly shaped olivine, plates of iron ore, a sparse, altered, glassy residuum and a little intersertal quartz. The rock is extremely fresh in comparison with members of the late-Carboniferous quartz-dolerite suite in this area and bears close resemblance to some of the Tertiary dykes described from north Ayrshire (Richey and others, 1930, pp. 306 – 308). RWE

CHAPTER 10

Structure of the post-Silurian rocks

INTRODUCTION

Within the district four main periods of earth-movement can be recognised, as follows:

1 **Caledonian Orogeny.** Pre-Lower Devonian (Lower Old Red Sandstone) earth-movements described in Chapter 2.
2 **Late-Caledonian Orogeny.** Post-Lower Devonian (Lower Old Red Sandstone), pre-Devono-Carboniferous (Upper Old Red Sandtone) earth-movements.
3 **Early Hercynian Orogeny.** Intra-Carboniferous earth-movements.
4 **Hercynian Orogeny.** Late-Carboniferous to early Permian earth-movements.

The principle structures, folds and faults affecting the post-Silurian rocks in the district are shown on Figures 19 and 20.

LATE-CALEDONIAN OROGENY

Deposition of the Lower Devonian rocks was brought to an end by the onset of earth-movements representing the final phase of the Caledonian Orogeny. The resulting unconformity between Lower and Upper Devonian strata, which is seen on the Berwickshire coast at Eyemouth, cannot be demonstrated in the present district. During this late-Caledonian orogenic phase a system of large faults, including the Lammermuir and the Dunbar–Gifford faults, was initiated or reactivated. These form part of the group of faults which together constitute the Southern Upland Fault-belt (Anderson, 1951, pp. 93–98). The direction of throw varies from place to place along the length of this complex of dislocations, and George (1960, p. 89) has postulated strong strike-slip movements.

The powerful Lammermuir Fault, trending NE to SW, is one of the most important structural features of the district. With a major downthrow to the north-west, this dislocation separates the closely plicated Ordovician and Silurian strata of the Lammermuir Hills from the more gently folded Devonian and Carboniferous rocks to the north-west. The position of the Lammermuir Fault in the Fala area has been revised from that shown on earlier maps on the basis of unpublished geophysical data which show a well-developed gravity-change, thought to be associated with the fault. The only exposure of the fault within the district is in the Birns Waters at Stobshiel where a soft, red and brown mottled clay-gouge separates Ordovician greywackes and shales to the south-east from Upper Old Red Sandstone sandstones and mudstones to the north-west.

The Dunbar–Gifford Fault runs parallel to the Lammermuir Fault, and about 2 to 2.5 km to the north-west of it. The downthrow is also to the north-west but of lesser magnitude, separating Devono-Carboniferous and lower Calciferous Sandstone Measures strata to the south-east from higher Calciferous Sandstone Measures strata to the north-west. The previously mapped position of the fault was confirmed at two points by geophysical traverses.

EARLY HERCYNIAN OROGENY

Movements occurring during the deposition of the Carboniferous rocks have been described by various authors including Anderson (1951), Francis (*in* Craig, 1965), Goodlet (1957; 1959), Kennedy (1958) and George (1960; *in* Craig, 1965) in the course of their studies of the history of sedimentation in the Midland Valley of Scotland. Evidence within the district for significant movement contemporaneous with sedimentation is provided by the unconformity at the base of the Passage Group which cuts out the Castlecary Limestone in the Cockenzie–Port Seton area (p. 39).

HERCYNIAN OROGENY

The principal structures of the Carboniferous rocks date from the climax of the Hercynian Orogeny in late-Carboniferous or early Permian times.

Knowledge of the structure varies greatly throughout the district. In the East Lothian Coalfield the nature of the folding and faulting is known in detail from information from boreholes and mine plans and from exposures in limestone quarries in the lower part of the Lower Limestone Group. The structure of the Devono-Carboniferous rocks and the Calciferous Sandstone Measures is less well known due to the cover of superficial deposits. Data derived almost entirely from natural exposures suggests that the pattern of folding and faulting resembles that found in the coalfield.

Eastern area – Calciferous Sandstone Measures

The Calciferous Sandstone Measures including the Garleton Hills Volcanic Rocks in the east of the district have a regional dip to the west. Along the north-east margin of the district, between North Berwick and the Dunbar–Gifford Fault, the lower volcanic rocks and the sediments are folded into a series of gentle anticlines and synclines with a NE–SW trend and SW-plunging axes. The volcanic rocks are preserved in the cores of the Whitekirk and Prestonkirk synclines and the underlying sediments outcrop in the Balgone, Crauchie and Traprain anticlines. The dome in the last of those, round Traprain Law, is in part at least produced by intrusion of the phonolite laccolith. There is little evidence of folding in the poorly-exposed central part of the district, though the gentle folding seen on the coast at Weaklaw Rocks (Plate 17) may be typical of structures inland.

Few faults are known from this area. The Gleghornie Fault and other small faults are on the NE–SW trend of the

Plate 17 Gently folded sedimentary rocks, Calciferous Sandstone Measures, Weaklaw Rocks, north of Dirleton (D3034)

major faults. The North Berwick, Ferny Ness and Alderston faults trending WNW–ESE are possibly related to the W–E faults in the coalfield. The North Berwick Fault is well exposed in the intertidal area west of the harbour, but the other two faults are conjectural. The Alderston Fault, with a southerly downthrow, was inserted mainly to account for the marked increase in the width of the outcrop of Calciferous Sandstone Measures sediments overlying the Garleton Hills Volcanic Rocks which takes place in that vicinity. This increase may be due in part to an unconformity at the top of the volcanic rocks . Faults with a NNW–SSE trend include the Archerfield and Wamphray faults, both with a downthrow to the west, and parallel to faults in the coalfield.

South of Dunbar–Gifford Fault

In the strip between the Dunbar–Gifford and Lammermuir faults the strata have a regional dip to the south-west. This dip is modified by open dome and basin structures with NW–SE axes which produce the Gifford Basin of the Calciferous Sandstone Measures and the Upper Old Red Sandstone inliers of Humbie and Costerton. Faulting is almost unknown apart from faults on the west side of the Humbie Inlier roughly parallel to the major faults, and the NW–SE-trending Garvald Mains Fault.

South of the Lammermuir Fault the Longyester Outlier and the Fala Outlier of Lower Old Red Sandstone strata are similar. The offset of the dome and basin structure across the faults suggests some lateral movement of the Dunbar–Gifford and Lammermuir faults.

Western area – East Lothian Coalfield

The two coalfield basins of the Midlothian Coalfield and the East Lothian Coalfield are separated by the irregular D'Arcy–Cousland Anticline which is cut obliquely by the Crossgatehall Fault (Figure 19). This important anticlinal structure has, in the past, been drilled for oil and gas (Tulloch and Walton, 1958, pp. 120–121). The Midlothian Basin is abruptly truncated to the north-west by the Pentland Fault, whereas the major Dunbar–Gifford Fault lies south-east of, but does not affect, the East Lothian Basin.

The overall structure of the East Lothian Coalfield is that of a complex open syncline, composed of gentle dome and basin structures, intersected by numerous minor faults (Figure 20). Dips are in general very low, except in the proximity of faults. The deepest of the basins form the Seton and Elphinstone outliers of Upper Limestone Group strata to the

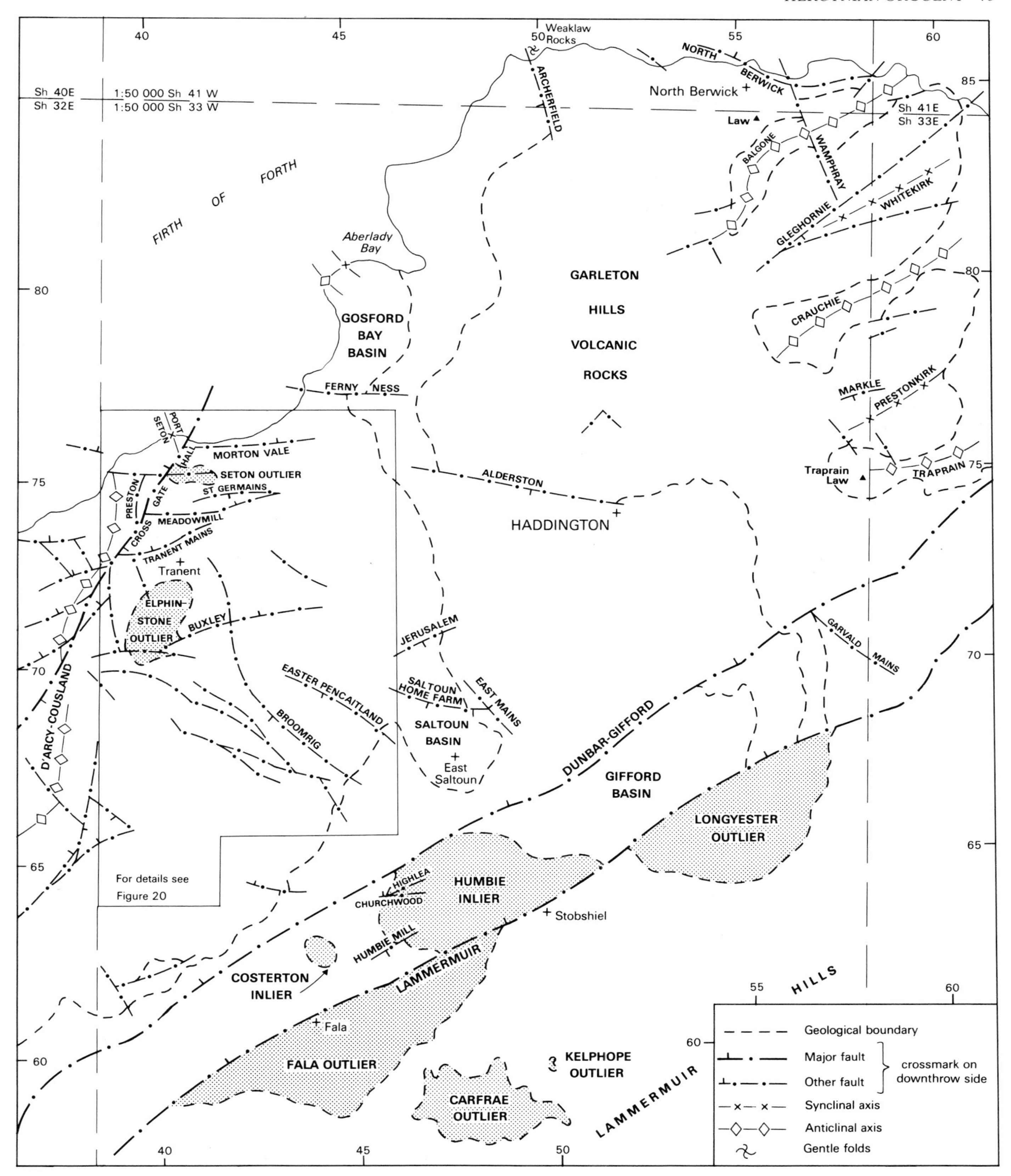

Figure 19 Main structural features in the post-Silurian rocks

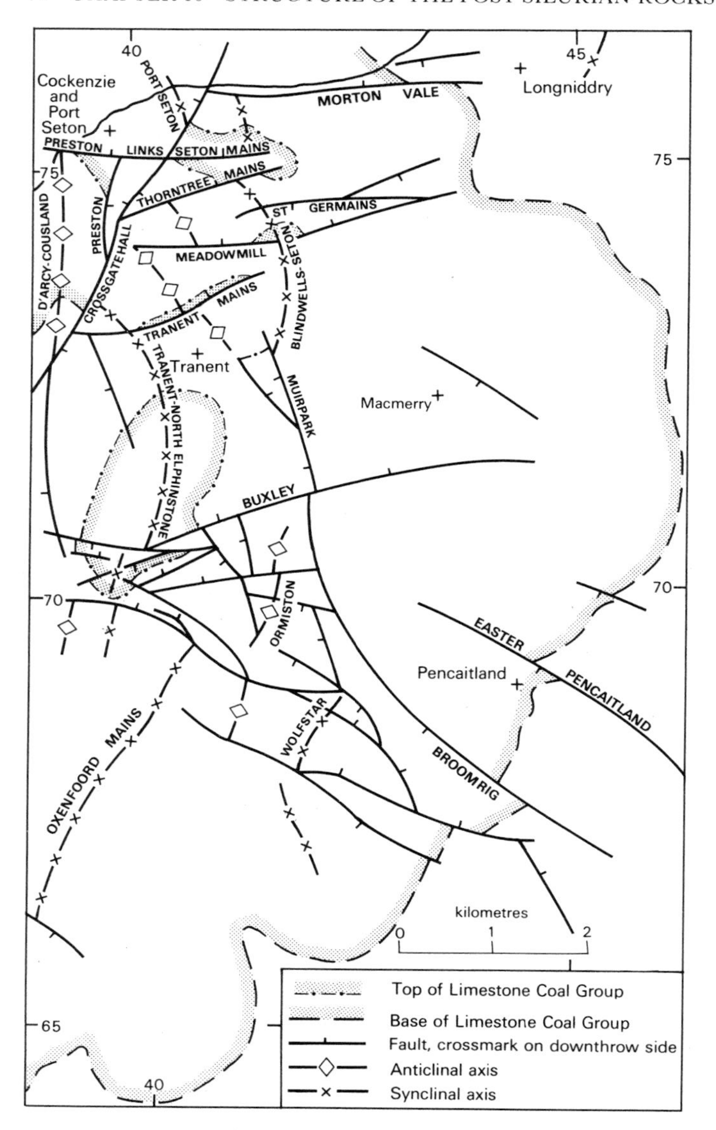

Figure 20 Structure of the East Lothian Coalfield

north and south of Tranent. The principal folds are truncated by the faults, so the folding probably took place before or during the faulting. Within the coalfield the fold-axes have a general NNW–SSE or N–S trend to the north of Tranent and further south the predominant axial trend is NE–SW (Figure 20). With open dome and basin structure, the axial plane is not well defined.

The north end of the D'Arcy–Cousland Anticline lies just within the western boundary of the district south of Cockenzie, where it trends N–S and is northward-pitching. The complementary Port Seton Syncline to the east plunges to the SSE. Both folds are truncated by the major Crossgatehall Fault. South and east of this fault there are three principal downfolds, namely the Tranent–North Elphinstone Syncline, the Blindwells–Seton Syncline, and the Oxenfoord Mains Syncline. The first two features form a series of basinal structures, truncated by W–E faults, and in both folds a SW axial trend in the south swings towards the NNW in the north part. The axis of the Oxenfoord Mains Syncline has a general NE trend, and the south-west part of the fold is asymmetrical with gently dipping beds on the eastern limb and dips of 25° to 30° on the western limb near the west margin of the district.

The Ormiston Anticline and the Wolfstar Syncline are less prominent structures, both with SW–NE axial trends, and both truncated by faults more or less at right angles to the fold axes.

Outwith the coalfield, folds with a NW or NNW axial trend occur in several localities. These include the Gosford Bay and East Saltoun basins and the minor anticline and syncline on the north limb of the latter basin south of Aberlady Bay. The East Saltoun Basin plunges to the north-west and produces a prominent south-easterly extension of the outcrop of the Lower Limestone Group. The Gosford Bay Basin also plunges to the north-west, but the minor folds south of Aberlady Bay plunge to the south-south-east.

The two most common fault-trends in the coalfield are approximately W–E and NW–SE. Several of the most important dislocations follow a W–E trend, with downthrow to the north. The principal ones are, from north to south (with approximate downthrow in brackets), the Morton Vale Fault (35 m), the Preston Links Fault (73 m), the Seton Mains Fault (46 m), the Meadowmill Fault (37 m), the Tranent Mains Fault (18 m) and the Buxley Fault (29 m). With the exception of the Preston Links Fault, the throw of these fractures generally decreases eastwards. The St Germains Fault, which also trends approximately W–E, has a southerly downthrow, and along with the easterly part of the Meadowmill Fault to the south, forms a graben structure.

One of the major dislocations in the coalfield area is the Crossgatehall Fault which runs in a south-south-west direction from the coast at Port Seton (Tulloch and Walton, 1958, pp. 121, 124). In the district the throw is down to the north-west along the northern part of the fault to the north-west of the Preston Fault, and to the south-east to the south of that fracture. At Port Seton and Prestonpans, apparent sinistral displacements of quartz-dolerite dykes by the Crossgatehall Fault (Figure 20) and possible sinistral displacement of the axis of the D'Arcy Cousland Anticline, just outwith the district, imply that movement along the Crossgatehall Fault has occurred after the formation of the W–E faults, considered contemporaneous with the emplacement of the dykes.

In the coalfield area one of the more important fractures with a NW trend is the Muirpark Fault. The dislocation has a westerly downthrow of as much as 73 m, and in the northern section the fault forms the eastern side of a southerly convergent graben containing an inlier of Upper Limestone Group strata.

WT, ADM

CHAPTER 11
Quaternary

INTRODUCTION

No sediments have been recorded in the district from the long time-interval between the Carboniferous and the Quaternary. Any deposits laid down during this period must have been removed by subsequent erosion.

It has been suggested that a marine-planed surface was formed over Scotland during the Middle Cretaceous. Uplift and gentle eastwards-tilting produced a surface on which the drainage pattern of eastern Scotland was initiated. The River Forth is one of the east-flowing consequent rivers and the River Tyne its tributary. Tertiary erosion cut down into the underlying rocks and the initial, simple, river pattern was modified as hard or soft rocks were encountered (Bremner, 1942; Linton, 1951; Sissons, 1967).

Glacial deposits have infilled and obscured the course of pre-glacial river valleys. Evidence from West Lothian and Edinburgh shows an extensive network of buried channels related to a sea-level in the Firth of Forth before the last main glaciation, lying well below present sea-level (Mitchell and Mykura, 1962). In the present district, a buried channel of the River Tyne probably underlies the broad boulder clay- and alluvium-covered plain from Ormiston through Haddington to Beanston, though sub-surface information is lacking. At Fala Dam a borehole proved at least 64 m of superficial deposits in a buried channel. At Beanston the River Tyne originally flowed north-east through the Garleton Hills ridge in a valley which is now blocked with till as suspected by Barrow (*in* Clough and others, 1910, p. 171) and since confirmed by geophysical evidence. The pre-Glacial River Tyne continued towards the Firth of Forth through the area now drained by the eastern Peffer Burn, but blockage of the valley at Beanston forced the post-Glacial river to cut a rock gorge eastwards to East Linton. The drift-filled valley running west–east along the western Peffer Burn may be another buried river channel, or may be a glacial gouge. Small, north-flowing, buried channels have been proved in the opencast site at Blindwells, near Tranent.

AGE OF THE GLACIATION AND PLEISTOCENE DEPOSITION

The Pleistocene period included several glacial and interglacial stages. Although these have been recognised in continental Europe and to some extent in England, in central Scotland the latest glaciation, the Devensian, obliterated evidence of all earlier glaciations and interglacials and deposited an extensive till. The Devensian glaciation, equated with the Wurm stage in the Alps and the Weichsel stage in North Germany (Flint, 1957, pp. 404–413), lasted from 80 000 to 10 300 years BP.

Warming of the climate and retreat of the Devensian ice-sheet began about 18 000 years BP. Readvance of the ice and cooling of the climate occurred around 15 000 years BP. During the late-Glacial period, which lasted till the ice finally cleared about 10 300 years BP, most of the glacial drainage channels were cut and the stratified glacial deposits of sand and gravel or lacustrine silt and clay were laid down, and the high late-Glacial raised beaches were formed.

The main Flandrian deposits comprise the post-Glacial raised beach and associated blown sand deposits, which formed during a period of static sea level between 7000 and 5000 years BP. Other post-Glacial deposits include peat, river terrace and flood-plain alluvium and lake deposits. Recent deposition is restricted to coastal beach and sand dunes and river flood-plains.

DIRECTION OF ICE-FLOW

The regional ice-flow in Central Scotland was interpreted in the 19th century using evidence from lineations and indicator stones. Snow and ice accumulating in the Southern Uplands around Broad Law, Hartfell and White Comb flowed to the north-east, encountering south-easterly flowing Highland ice from the Loch Lomond area. The merged ice flowed almost due eastwards across the Lothians, being deflected to the east-north-east by the Lammermuir Hills.

Passage of the ice-sheet moulded the underlying rocks and the ground moraine or till which it deposited. The moving ice-sheet, acting as a selective erosion agent, left the hard igneous rocks as upstanding crags and cut deep hollows into the sedimentary rocks. Ice-moulding features are particularly prominent in the northern part of the district (Figure 21). Glacially striated rock surfaces are uncommon in the district. They are restricted to the craggy lava terrain of the Garleton Hills and to the area from Kingston Hill to Whitekirk Hill, where striations are preserved on hard rock surfaces recently stripped of the till. Crag and tail features are common in the areas of volcanic rocks. The most spectacular are formed by volcanic plugs like North Berwick Law which has a tail over 1 km long. Tail-features occur in the lee of volcanic plateaux such as Kingston Hill, and even small rock crags only a few metres high can have tail-features extending for a kilometre or more. Fluted till has formed away from the effects of crags, as in the ridged ground east of Kingston Hill and east of the Garleton Hills. To the south-east of Haddington large drumlinoid ridges up to 2 km long have been formed. In places, glacial drainage and other later erosion has modified the lineations. Lineations give a more consistent and more accurate indication of flow-direction than glacial striae, as the basal ice is easily deflected by irregularities in the rock surface (Table 4). The direction of ice-flow was almost due east on the western margin of the district, deflecting almost to east-north-east in the eastern parts of the district.

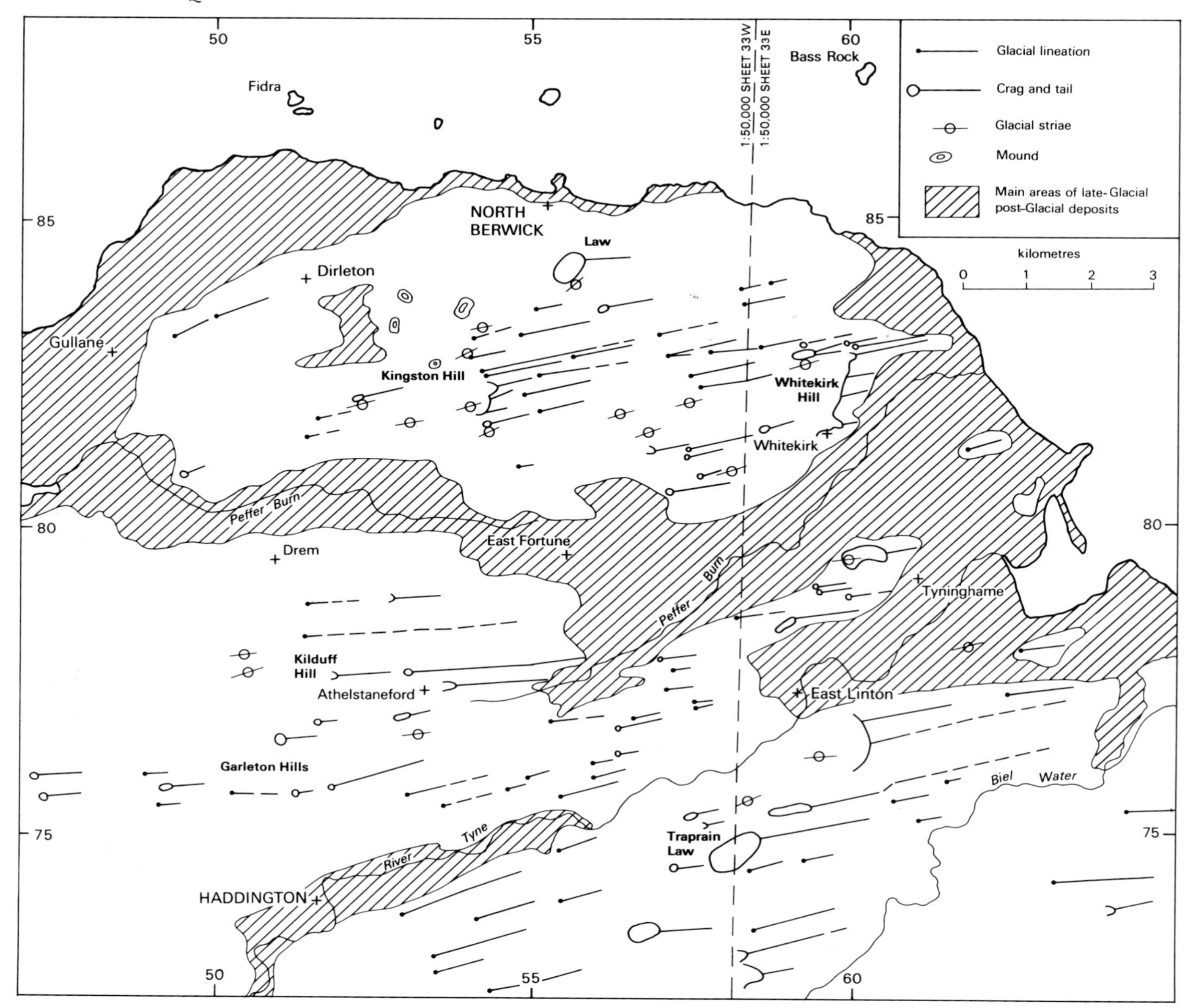

Figure 21 Ice-moulding in the northern part of East Lothian

Table 4 Comparison of lineations and directions of striae

	Linations	Striae
North Berwick Law	87°	45°
Kingston Hill	75° to 82°	65° to 80°
Whitekirk Hill	75° to 80°	70° to 80°
Kilduff Hill	84° to 87°	70° to 80°
Garleton Hills (west)	86° to 90°	nil
Garleton Hills (east)	74° to 84°	79°
SE of Haddington	68° to 80°	nil

GLACIOTECTONIC FOLDS

Movement and pressure caused by the ice-sheet travelling from west to east has pushed up and folded the top few tens of metres of Limestone Coal Group sedimentary strata just east of Tranent. Exploration boreholes in the NCB opencast site at Blindwells have shown two disturbed zones crossing the site, which were first interpreted as fault-zones. Excavations during development of the site have exposed glaciotectonic folds, rucks and thrusts, affecting only the upper part of the strata, and limited downwards by a plane of decollement. Below this plane the strata are undisturbed, showing that the structures are not allied to the many Hercynian folds and faults of the district.

DETAILS

The Limestone Coal Group strata at Blindwells lie in a gentle basin with a north–south axis; dips are generally flat to 5°, steepening to 10° in the east. The westerly disturbed zone is about 800 m long

Plate 18 Glaciotectonic folds in Diver Coal, Limestone Coal Group, Blindwells Opencast Site, Tranent (D3491)

and affects some 25 m of strata including the Great Seam, Diver and Tranent Splint coals. There is a plane of decollement in the strata above the Parrot Rough Coal. The strike varies from N10°W in the south to N10°E in the north. In the north the fold is a simple, plastic anticline overturned to the west whereas in the south there are complex folds with three anticlines particularly well brought out by the thin Diver Coal (Plate 18). Minor folds, thrusts, thinning of the strata on the limbs and thickening on the crests all occur on the folds. The thrusts are overturned consistently to the east, indicating pressure from the west. Overturning of the simple fold to the west suggests the pressure was transmitted at the base of the fold and the base was thrust under the top of the fold.

The easterly disturbed zone has not been so fully exposed. It is at least 1200 m long, strike varying from N20°E in the south to N10°W in the north, and effects at least 35 m of strata including the Great Seam, Diver, Tranent Splint and Parrot Rough coals. The plane of decollement has not been seen, but from borehole evidence it must be above the Lower Craw Coal which is structurally unaffected and displays an even, westerly dip.

The superficial deposits overlying the Limestone Coal Group strata are normally of till, a few metres thick. Buried channels, about 20 m deep, containing laminated clay and sand as well as till, cross the site. Their position just east of the two disturbed zones suggests that the presence of buried channels was a factor in the formation of the folding.

These unusual structures must have been the results of a rare combination of factors, including buried channels, a good slip-horizon and the state of the strata.

TILL (BOULDER CLAY)

Till mantles most of the low lying areas of Upper Palaeozoic rocks north of the Lammermuir Fault. Over much of this area it is commonly 5 to 10 m thick. In areas of volcanic rocks the till is thin and patchy, while thick accumulations are recorded from the buried channel at Beanston and around Fala. In the upland area of Lower Palaeozoic rocks, it only occurs sporadically in some valleys.

As most of the material is locally derived, the colour, stone-content and matrix of the till correspond closely to the nature of the underlying rocks. Erratic stones such as Highland schists, though significant in understanding ice-flow directions, are present in only small proportions. In the western half of the district the till was derived from Carboniferous sedimentary rocks and consists of dark grey, silty clay with dominantly sandstone pebbles and small proportions of limestone, ironstone, coal and other sedimentary rocks as well as mainly intrusive igneous pebbles. In areas of volcanic rocks in the north-east of the district around Kingston and Garleton Hills, the till is a mottled grey, purple and orange, silty or sandy with basalt, trachyte, dolerite and other igneous pebbles mixed with pebbles from the Carboniferous sedimentary rocks. In areas of Devono-Carboniferous rocks the till is red or purple, sandy clay with dominantly red sandstone pebbles and greywacke and quartzite cobbles derived from conglomerates. In uplands of Lower Palaeozoic rocks there is a brown sandy till with fragments almost entirely of greywacke, siltstone and shale. Weathering has commonly altered the top 1 to 2 m to a brown less-consolidated till.

Intercalations of sand and gravel in the till are seen in several stream sections in the Keith Marischal – Fala area. At Red Scar [438 631] on the east bank of Keith Water, two tills can be distinguished and in another section 1 km downstream [448 637] a third till occurs at the top of the section (Kendall and Bailey, 1908; Kirby, 1968; Sissons, 1974). Kirby (1968) equated the three tills to the succession seen in the Esk valley around Roslin; from the stone-content he suggested a westerly derivation for the Basal Till and the upper Roslin Till and a southerly derivation for the Intermediate Till as it has a high content of greywacke from the Southern Uplands.

Other records of stratified deposits within the till in the Fala area include: East Water west of Woodcote Mains [450 615]—3 m of brown-grey clay with sand laminae; Johnstounburn Water, above and below the House [460 615]—similar clay to above; Partridge Burn [431 599]—2 m of stiff grey clay with few pebbles and 5.4 m of sand and gravel with coal fragments above and thin clay bands below.

LARGE GLACIAL ERRATICS

Several large masses of shattered limestone and calcareous mudstone in the southern part of the district are regarded as large glacial erratics. The most notable, and the largest known glacial erratic in Scotland (Sissons *in* Craig and Duff, 1975, p. 142), lies just north of Kidlaw [509 642]. It measures 500 m from NW to SE, 400 m from NE to SW and has been extensively quarried. Limestone erratics of smaller size occur at Marl Law Wood [416 623] north-east of Crichton Dean, and west of Woodcote Mains at Meikle Law [452 605]. The evidence for considering these masses as glacial erratics rather than as outliers are (Kendall and Bailey, 1908; Bailey *in* Clough and others, 1910):

1 The limestones lie on Devono-Carboniferous or Calciferous Sandstone Measures strata and these relationships would require large unconformities not known elsewhere in the district.

2 The masses are more or less shattered, as could result from glacial transport, but not by faulting.

3 There is a great abundance of smaller limestone erratics in the boulder clay of adjacent areas.

It is likely that large scarps of limestone existed in the preglacial landscape, furnishing an abundant supply of limestone blocks and affording favourable opportunities for detachment and transport of large masses.

Smaller angular erratics of Carboniferous limestone and sandstone incorporated in reddish till are exposed in several sections along the East Water, near Woodcote Mains [451 610].

Evidence exists for a large erratic containing coal seams near Fala Dam [429 616]. Coal seams near the surface were worked in the past on the east bank of Fala Dam Burn. However, borehole evidence indicates superficial deposits in excess of 64 m infilling a deep buried channel, with a 7-m thick bed of sand at the top, containing blocks of coal.

DEGLACIATION

As the climate improved, the ice-sheet covering the district began to melt and break up. Highland ice retreated northwards towards the Firth of Forth, and Southern Upland ice retreated south to form an isolated ice-cap with valley glaciers. Ice-free ground appeared on the slopes of the Lammermuir Hills at successively lower levels as the ice-margin gradually receded. The large volumes of glacial meltwater released by the melting ice cut numerous glacial meltwater channels and laid down stratified glacial deposits mainly of sand and gravel (Figure 22). Ice lying in the lower ground blocked most of the river valleys, damming back the meltwater which overflowed and cut new channels along the ice-margin and across cols into other valleys. Channels vary greatly in length and depth from short ones only a few metres deep, to some several kilometres long and up to 30 m deep. Many channels were abandoned as dry valleys, while others are still utilised by the modern drainage. The glacial drainage flowed generally to the north-east parallel to the slopes of the Lammermuir Hills and down the gradient of the ice-sheet. The majority of channels are contour-channels cut by meltwater flowing between the ice-margin and the hills (Plate 19), while some meltwater escaped under the ice to lower ground flowing down glacial chutes perpendicular to the contours (Plate 20). The channels described below are indicated by numbers in Figure 22.

DETAILS

Lammermuir Hills

The few channels that have been cut into rock in the upland areas are generally isolated glacial chutes rather than contour-channels. Meltwater dammed up in the Gala Water valley, overflowed into the Tyne basin cutting the channel at the Cakemuir Burn (1) and probably also that at Crichton Dean (2). At an early stage of deglaciation, channels were cut into rock high up the Lammermuir Hills allowing drainage across cols at Wanside (3), Lammer Loch (4) and Dod Law (5), in each case forming prominent ridges of coarse greywacke gravel, particularly at Latch (4).

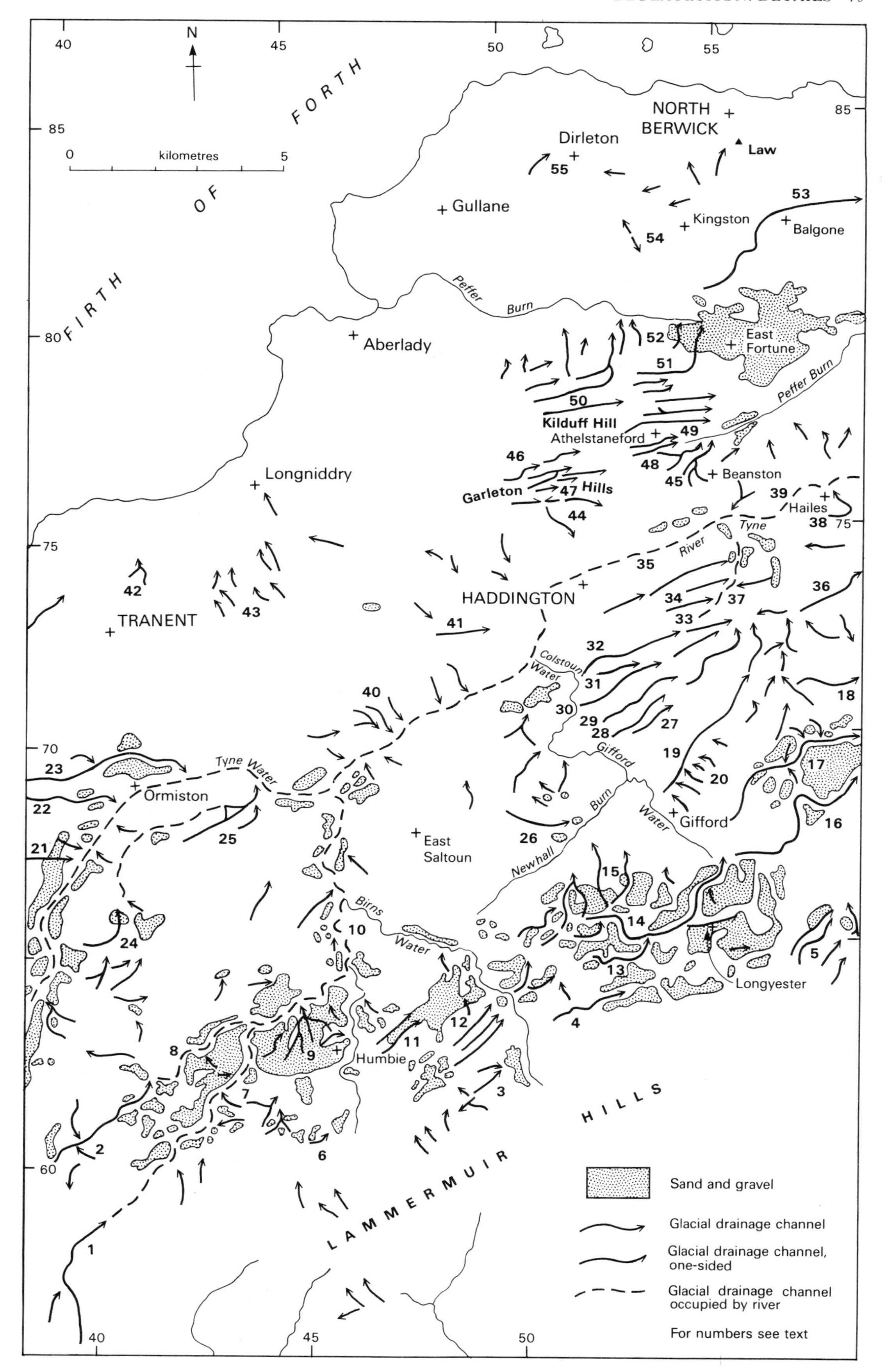

Figure 22
Glacial drainage and meltwater deposits

Lammermuir foothills

At a later stage, a standstill of the ice margin produced considerable erosion of sinuous, glacial-meltwater channels and deposition of extensive spreads of sand and gravel all along the foothills area. In the south-west, glacial water, possibly including overflow from both the Gala Water and Esk valleys, flowed initially at Woodcote (6) and later along the Fala Dam Burn (7), Salters Burn (8) and Keith Water (9). Thick deposits at Keith, mainly of sand, are cut by numerous large tributary channels (10). Further north-east a similar complex of channels and sand and gravel deposits extends from Humbie east to beyond Long Yester. Channels at Leaston (11) and Stobshiel (12) lead into large alluvium-bottomed channels at Long Newton (13) and Skedsbush (14). These flow east towards the Gifford Water while meltwater escaping north cut channels downslope towards the Newhall Burn (15). Unable to escape down the Gifford Water still blocked by ice, meltwater cut deep channels eastwards at Danskine (16), Bara (17), Chesters (18) and Morham (19). This last was fed by a prominent set of downhill channels (20) north-east of Gifford.

Esk Valley overflow

Glacial meltwater trapped in the upper reaches of the South Esk valley, overflowed through the Borthwick Gap into the Tyne Water above Pathhead and later crossed the watershed further north along three successive channels (21 – 23) south-west and west of Ormiston (Mitchell and Mykura, 1962, p. 117). All these are associated with extensive sand and gravel deposits along the Tyne Valley.

Area south of River Tyne

In the undulating till-covered area between the Tyne Water and the Birns Water, local standstills of the ice produced isolated channels and networks of channels, as at Dodridge (24), Fountainhall (25) and, further east, at Bolton Muir Wood (26). East of the Colstoun Water, at least nine successive ice-levels (27 – 35) are marked by distinctive open, one-sided channels, all with wide alluvial flats. These channels have only slightly modified the fluted-till topography, against which the ice was banked. Escape chutes lead from one channel to the next lower, and much of the meltwater drained into the deep channels south of Standingstone (36) and the channel now occupied by the Bearford Burn (37). A long, lower, esker-mound marks the position of a downhill flow of meltwater just east of the Bearford Burn. At Hailes a dry valley (38) marks an early abortive attempt to cut the Tyne Gorge (39), which was subsequently made by the Tyne drainage following glaciation, the former valley at Beanston having been blocked by till.

Between the River Tyne and the Peffer Burn

To the west, where much of the ground is undulating till-covered terrain, there are only a few local clusters of glacial channels, as at Jerusalem (40) and Letham (41) leading into the Tyne, and at Rig-gonhead (42) and Greendykes (43) on the slopes down to the Firth of Forth. At various stages, ice was banked up on both south and north slopes of the Garleton Hills and numerous channels were cut, but there was little deposition (Sissons, 1958). On the southern slopes there is one main marginal channel (44) and several

Plate 19 Glacial marginal channel, along contours, Springfield Wood (D3019)

Plate 20 Subglacial chute, to lower ground, Blegbie (D3018)

downslope chutes. Further east at Beanston, meltwater from the Tyne overflowed the ridge into the Peffer valleys, cutting a network of channels (45) into the till of the buried channel. They were not sufficiently deep, however, to allow the River Tyne to reoccupy its former valley and instead it cut a new gorge to East Linton. In the centre of the Garleton Hills, meltwater has followed glacial gouges (46), while other channels utilise the trap features of the lavas (47). Drainage is eastwards by channels at Cogtail Burn (48) and north of Athelstaneford (49), towards raised sea-levels in the east Peffer valley. On the north side of Kilduff Hill east-flowing contour channels mark successive retreat stages of the ice-sheet, as at Chesters (50) and West Fortune (51), and lead ultimately into downslope channels (52) which feed into the raised sea-level of the west Peffer Burn and the moundy sand and gravel deposits of East Fortune.

North of the Peffer valleys

A large and sinuous channel leads to the north-east at Balgone (53). Its south-east side is formed by cliffs of basalt lavas, its north-west side is smoothed till and its base is a wide alluvial strip. This channel marks the site of an ice-margin coincident with the edge of the rock, and it carried overflow from the west Peffer Burn. A network of channels (54) at Kingston allowed meltwater to cross the watershed. An isolated channel (55) at Dirleton is similar but smaller than the Balgone channel, while there are also a few small channels south of North Berwick.

GLACIAL AND FLUVIOGLACIAL SAND AND GRAVEL

Stratified deposits of sand and gravel were laid down by meltwater as the ice-sheet melted and the ice-front retreated northwards. Glacial deposits commonly form hummocky topography, kames and kettles, mounds and esker-ridges. Fluvioglacial deposits form valley-side terraces. Distinction between the two is blurred and has not been attempted in the district. Several stages of glacial retreat occurred, each with various glacial and fluvioglacial deposits, and later episodes caused dissection of previous deposits.

Four main areas of deposition occur; along the valley of the Tyne Water above Ormiston, along the valleys of the Birns Water–Keith Water–Salters Burn, along the Lammermuir foothills from Leaston to Carfrae, and in the north around East Fortune. Geological surveys for sand and gravel subsequent to the publication of the 1:50 000 drift map (McAdam, 1983) have increased understanding of the areas of deposits.

Rapid variation in grain-size and lithological content occurs throughout the deposits (McAdam, 1978). High ridges, mounds and eskers along the Lammermuir foothills are commonly formed of coarse boulder-gravel with little sand, passing to varied forms of deposits on the lower ground containing sand and fine gravel. The bulk of the finer material, silt and clay, was washed into glacial lakes or out into the late-Glacial sea. Greywacke forms most of the pebbles in deposits along the Lammermuir foothills, the remainder being mainly red sandstone pebbles. The proportion of Carboniferous rocks, mainly red, brown and white sandstones locally with igneous rocks, increases rapidly northwards, and these form over half the pebbles around Ormiston and Pencaitland. In the moundy deposits around East Fortune, modified by action of the late-Glacial sea, the gravels have a mixed pebble content of Carboniferous sandstones and other sediments, various igneous rocks, and greywacke.

CHANGES IN SEA-LEVEL, LATE-GLACIAL TO PRESENT

Sea-levels have varied considerably during the last deglaciation and subsequently (Donner, 1963; Sissons, 1967; Goodlet, 1970). The variation was caused by interaction of two major factors; firstly the lowering of world sea-level due to locking up of water in ice-caps, and secondly isostatic depression of the land, caused by the weight of the ice, followed by uplift resulting from the disappearance of the ice. Because the effect of isostatic depression was generally dominant, sea-levels were higher than at present, resulting in raised-beach features and deposits. The weight of ice was greatest in the centre of Scotland, so that subsequent isostatic uplift was greatest there, and the older beaches tilt away from the centre (Donner, 1963). The oldest recorded beaches are late-Glacial, formed 16 000 to 13 000 years BP when sea-level in this area lay between 20 and 30 m OD. Examples of beaches of this age are found in the upper parts of the valleys of both the Peffer Burn (west) and the Peffer Burn (east) and in patches from Port Seton to Drem. During the period 12 000 to 10 000 years BP, evidence from other areas indicates a sea-level well below the present one. From 7000 to 5000 years BP sea-level was constant at about 8 m OD and the resulting Main Postglacial Shoreline is seen as a narrow raised-beach along much of the coast and in estuarine flats in Aberlady Bay and at Dirleton. Sea-level subsequently fell gradually to its present position.

LATE-GLACIAL RAISED BEACH

DETAILS

Prestonpans to Drem

The eastern end of the extensive, high, raised-beach terrace which runs eastwards from Edinburgh is present in the Prestonpans area. The beach can be traced eastwards by Longniddry and Drem as far as West Fortune as a well-developed terrace, with a back-feature which falls from about 30 m OD at Prestonpans to about 25 m OD at Drem. Although some beach deposits are present at Prestonpans, to the east only a patchy, thin, beach deposit lies on a till-covered terrace. South of Port Seton another raised-beach, back-feature has been identified at a level of about 15 m OD.

Peffer Burn (West)

Estuarine, alluvial deposits cover the bottom of the W – E running valley of the Peffer Burn in a strip 5 km long by about 1 km wide. The deposits all lie below 15 m OD, are banked up against till to the north and south, against glacial sand and gravel deposits to the east, and are overlain by deposits of the post-Glacial raised beach 8 m OD to the west. On the north side of the valley the deposits form a distinct sloping flat with a well-marked back-feature cut into the till. On the south side the features are less clearly defined. The deposit is characterised by small side-channels flowing towards the Peffer Burn cut mainly within the area of the deposit. Natural sections are lacking but evidence from trenches and auger holes along the valley sides show the deposit is at least 3 m thick and consists of dark to light grey, laminated clay, silty in parts, with isolated pebbles. Along the centre of the valley are moundy areas of fine silt and sand. At Muirton an auger hole [5169 8026] proved a pale greenish-grey, diatomaceous silt with numerous small freshwater bivalves and gastropods, indicating the former presence of a lake.

Gullane

Fragments of raised-beach back-features are cut into the till at various levels to the east of Gullane. Patchy deposits may be associated with the features.

North Berwick

Two small areas of raised-beach deposits on top of the cliffs east of North Berwick form irregular sloping terraces with back-features at about 22 m OD. There is a sandy beach deposit with some blown sand.

Peffer Burn (East)

The western part of the eastuarine, alluvial deposits which cover the bottom of the valley of the eastern Peffer Burn lie in the district. The deposits occur in a strip about 1 km wide and lie below 22 m OD. They are banked up against till to the south, but to the north and west are separated by moundy glacial sand and gravel deposits from the estuarine alluvial deposits in the west Peffer Burn. The deposit blankets the pre-existing topography without forming any flats, shore-lines or back-features. Characteristic are small side-channels 1 to 6 m deep flowing towards the Peffer Burn, usually cut within the area of the deposit, though some start higher up the slope. The deposit is rarely exposed in natural sections, but it forms a grey or yellowish, stoneless, clay soil, in places becoming silty or finely sandy. A traverse line of shallow auger holes from Jagg to Janefield proved deposits of grey-blue and purple, laminated clay to silty clay, with sand and fine gravel up to 6 m thick.

POST-GLACIAL RAISED BEACHES

DETAILS

Prestonpans to Aberlady

Along this stretch the well-developed post-Glacial raised beach forms a narrow bench backed by a prominent back-feature commonly at 8 m OD. This feature is obscured by dunes of blown sand particularly between Longniddry and Craigielaw Point. The lithology and fauna of the beach is seen in several good sections eroded at H.W.M., which show it to consist of shells, sand and shingle (Plate 21).

Plate 21 Raised beach deposit, post-Glacial, mainly oyster shells, Gosford Bay (D3260)

Luffness

An extensive raised estuarine-flat was formed east of Luffness in post-Glacial times similar to the present-day estuary of the Peffer Burn in Aberlady Bay. The deposits are bounded by till to the north-east and the south, cut late-Glacial estuarine deposits to the east and are obscured by blown sand to the north-west. The deposits are poorly exposed, but the soils developed indicate that they consist of silt, clay and fine sand, similar to the mud and sand flats at present in Aberlady Bay. The Luffness Cocklebed has a typical estuarine fauna dominated by cockles.

Dirleton

North of Dirleton an extensive sloping flat of raised-beach deposits extends up to a poor back-feature at about 15 m OD. This height suggests the beach is partly late-Glacial. The beach deposits appear from the nature of the soil to be mainly sandy.

North Berwick

The lower part of the town of North Berwick is built on a raised-beach terrace with a fine cliff-feature behind. The deposits are likely to be similar to the present-day beach, mainly sand and shingle resting on rock. ADM

PALAEONTOLOGY

Pleistocene faunas were recovered from two boreholes. The first, near Fenton Barns [5101 8069], revealed clays apparently of late-Devensian age. Samples collected between 1.75 and 2.75 m below surface level contained a shallow-water, marine microfauna of restricted diversity. The fauna, which was identified by I. P. Wilkinson, was dominated, almost to the exclusion of other species, by the foraminifer *Elphidium clavatum* Cushman. Other foraminifera present were *Guttulina* cf. *glacialis* (Cushman and Ozawa), *Lagena gracillima* Seguenza, *Protelphidium orbiculare* (Brady) and *Pseudopolymorphina novangliae* (Cushman). This assemblage is indicative of a harsh, but not necessarily fully arctic, environment. Cold water marine clay was also proved south-east of Chapel [5329 8067] where borehole samples collected between 2 and 3 m depth contained assemblages essentially similar to the above. At Chapel, however, the faunal dominance is shared between *E. clavatum* and *E. asklundi* Brotzen.

Along the coast from Longniddry to the eastern margin of the sheet, intermittent exposures of raised beach, representing the Main Postglacial Shoreline, have provided post-Glacial macrofaunas, a number of which are listed in Table 5.

Table 5 Post-Glacial fossils

As a rough guide to relative abundance, taxa occurring singly, or in very small numbers, are indicated by the letter P (present). Those which are relatively common are indicated by C (common) and those occurring in large numbers are shown as A (abundant).

Locality	1	2	3	4	5	6
Gastropoda						
Acmaea sp.	.	.	P	.	.	C
Alvania crassa (Kanmacher)	.	.	.	.	.	P
A. punctura (Montagu)	.	.	.	.	.	P
Buccinum undatum Linné	P	P	.	.	.	.
Crysallida indistincta (Montagu)	.	.	.	.	.	?
C. obtusa (Brown)	.	.	.	.	.	P
C. spiralis (Montagu)	.	.	P	.	.	.
Cingula semistriata (Montagu)	.	.	.	.	.	P
Cylichna cylindracea (Pennant)	.	P	.	.	.	.
Cytharella smithi? (Forbes)	.	.	.	.	.	P
Emarginula reticulata Sowerby	.	.	.	C	.	.
Gibbula cineraria (Linné)	C	C	P	.	P	P
Hydrobia ulvae (Pennant)	.	.	C	A	.	C
Lucuna vincta (Montagu)	.	.	.	P	.	P
Littorina littoralis (Linné)	A	A	A	P	P	P
L. littorea (Linné)	C	C	C	P	P	.
L. saxatilis (Olivi)	A	A	A	P	.	C
Mangelia coarctata (Forbes)	.	P	.	.	.	P
Nassarius incrassatus (Ström)	P	C	.	P	P	C
Nucella lapillus (Linné)	C	C	P	P	P	P
Odostomia unidentata (Montagu)	.	P	P	.	.	?
Onoba semicostata (Montagu)	P	C	P	.	.	C
Patella vulgata (Linné)	A	C	C	P	C	P
Patina pellucida (Linné)	.	P	.	.	.	P
Retusa obtusa (Montagu)	.	.	.	C	.	.
Rissoa parva interrupta Adams	P	C	C	P	.	A
Rissoa parva (da Costa) ss.	.	.	.	P	.	P
Turritella communis Risso	C	P	.	P	P	P
Velutina velutina (Müller)	P	.	.	.	.	.
Bivalvia						
Abra alba (Wood)	.	.	.	.	.	P
A. prismatica (Montagu)	.	.	.	.	.	P
Acanthocardia echinata (Linné)	P	.	.	.	.	.
Arctica islandica (Linné)	C	.	.	.	P	.
Cerastoderma edule (Linné)	.	.	P	A	C	C
Chlamys distorta (da Costa)	.	.	.	P	.	.
C. opercularis (Linné)	.	P	.	.	.	P
C. varia (Linné)	.	.	.	.	.	P
Chlamys sp.	P	P	P	.	.	.
Corbula gibba (Olivi)	P	P	P	C	P	C
Donax vittatus (da Costa)	.	.	.	.	P	C
Dosinia exoleta (Linné)	?	.	.	P	P	P
Ensis siliqua (Linné)	.	.	.	.	A	.
E. sp.	.	.	.	P	.	.
Heteranomia squamula aculeata (Müller)	.	.	.	.	.	C
H. squamula squamula (Linné)	.	.	P	P	.	C
anomiid	.	.	.	P	.	.
Hiatella arctica (Linné)	.	.	P	P	.	P
Lasaea rubra (Montagu)	.	.	P	P	.	.
Lutraria lutraria (Linné)	C	.	.	C	A	.
Macoma balthica (Linné)	.	.	.	C	.	C
Monia patelliformis (Linné)	P	.	.	.	.	.
Mya truncata (Linné)	.	.	.	.	P	.
Mysella bidentata (Montagu)	.	P	C	C	.	P
Mytilus edulis Linné	.	P	P	.	P	C
Nucula sulcata Bronn	.	.	.	.	.	P
N. sp.	.	.	.	.	.	P
Ostrea edulis Linné	C	A	C	.	.	.
Parvicardium ovale (Sowerby)	.	.	.	P	.	C
P. scabrum (Philippi)	.	C	.	P	.	.
pholadacean fragments	.	.	.	.	.	P
Saxicavella jeffreysi Winckworth	.	.	.	P	.	.
Spisula elliptica (Brown)	.	.	.	C	P	C
S. solida (Linné)	.	.	.	.	P	P
S. subtruncata (da Costa)	P	C	.	C	P	P
Tellina (Fabulina) fabula Gmelin	.	.	.	C	.	.
T. tenuis da Costa	.	.	.	P	.	C
T. sp.	.	.	.	P	P	.
Thracia villosiuscula (Macgillivray)	.	.	.	P	.	.
T. sp.	.	P	.	.	.	.
Turtonia minuta (Fabricius)	.	.	.	P	.	.
Velutina velutina (Müller)	.	P	.	.	.	.
Venerupis sp.	.	.	.	P	.	
Venus striatula (da Costa)	P	.	.	C	A	A

Localities

1 Longniddry shore [4398 7764] 0.25 m of shells and pebbles.
2 Gosford Bay [4435 7780] 0.6 m of boulders and pebbles with conspicuous oysters and smaller shells on 0.9 m of finer sediment.
3 Kilspindie shore [4512 8028] 0.75 m of shelly sand and pebbles.
4 Luffness [469 807] shells obtained from rabbits' burrows.
5 Gullane Point [4628 8302] 0.2 m of coarse shelly sand and pebbles.
6 Craigielaw [4467 7980] 0.2 cm of stratified blown sand.

Examination of present day beaches along the Forth Estuary shows that there is considerable faunal variation from one location to another, areas in which numerous rocks outcrop supporting different assemblages from sandy beaches or tidal mudflats. Other factors such as salinity levels or exposure also affect faunal distribution. Understandably, therefore, the raised-beach faunas also vary. At Longniddry and Kilspindie (localities 1–3) numerous rock outcrops occur and the common occurrence of rock-dwelling gastropods such as *Littorina spp.* and *Patella vulgata* in the raised beach deposits indicate that those beaches were equally rocky during the early post-Glacial period. Interpretation of post-mortem beach assemblages must necessarily be tentative due to the extent to which some appear to have been reworked. No microfossils were found in any of the sediment samples examined. This was possibly due to leaching, a factor which, along with physical abrasion, could also be responsible for the elimination of some of the more delicate molluscan taxa and the distortion of the perceived faunal composition.

The fauna from a temporary section at Luffness (Locality 4) was described by Smith (1972, pp. 31–49). The sediments were shown by radiocarbon dating to have been deposited around 5500 years BP. The deposit, known as the 'Luffness Cockle Bed', is not presently exposed, although shells can be found in areas disturbed by rabbit burrowing. As the name implies, the faunal assemblage is dominated by the bivalve *Cerastoderma edule*. The fauna, which also includes other intertidal bivalves such as *Macoma balthica* and the hyposaline gastropod *Hydrobia ulvae*, reflects deposition in conditions similar to the tidal sandy mudflat presently adjacent to the deposit.

Between Gullane Point and Ironstone Cove (Locality 5) the fauna is dominated by burrowing bivalves such as *Ensis siliqua* and *Lutraria lutraria* with *Cerastoderma edule* and *Venus striatula* also common. Although *P. vulgata* is present in significant numbers the assemblage on the whole suggests

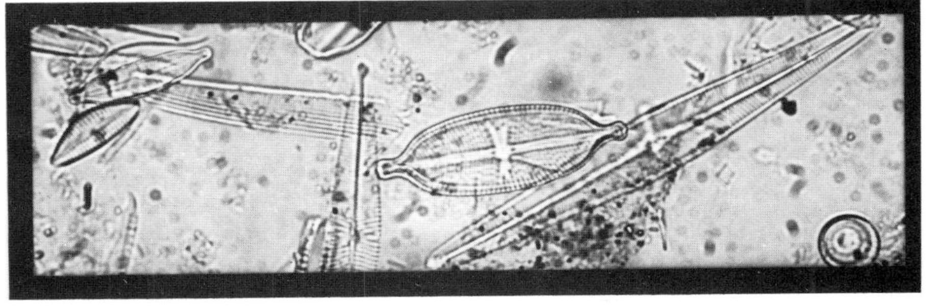

Plate 22 Quaternary fossils.

Figures 1-5, freshwater molluscs from a borehole near Muirton Farm (p. 86)

1 *Armiger crista* × 12
2 *Lymnaea peregra* × 6
3 *Gyraulus laevis* × 8
4 *Sphaerium corneum* × 5
5 *Pisidium nitidum* × 5

Figures 6–9, marine molluscs from raised beach at Longniddry (locality 1, p. 84)

6 *Patella vulgata* × 15
7 *Nucella lapillus* × 1
8 *Venus striatula* × 1.5
9 *Littorina littorea* × 1.5

10 Diatoms from same locality as Figure 1 × c. 700

that the beach was, as now, essentially sandy.

About 2 m of stratified blown sand, of uncertain age, is exposed south of Craigielaw (Locality 6). Along with a high proportion of comminuted shell debris, this deposit contains a diverse molluscan fauna characteristic of the sandy beach from which it was apparently derived.

At Muirton Farm [5169 8025] a shallow borehole revealed a diatomaceous lake-marl containing a restricted, but apparently temperate; fauna. Samples collected between 1.01 and 2.74 m below surface level included the gastropods *Armiger crista* (Linné), *Gyraulus albus* Müller, *G. laevis* (Alder) and *Lymnaea peregra* Müller, and the bivalves *Pisidium millium* Held, *P. nitidum* Jenyns and *Sphaerium corneum* (Linné). Freshwater ostracods were also present in the deposit and were identified by Dr J. R. Haynes and Mr P. Sherrington (University College of Wales) as *Candona candida* (Müller), *Cypria* cf. *ophthalmica* Jurine, *Erpetocypris reptans* (Baird), *Eucypris sp.* and *Limnocythere inopinata* (Baird). DKG

PEAT

Immediately following glaciation, conditions encouraged extensive colonisation by birch-hazel-pine and heather-sphagnam floras and initiated the development of hill-peat on till and rock in the Lammermuir Hills. This covering has been dissected by erosion and human stripping. The largest extant area is on Fala Moor, regarded by Ragg and Futty (1967, pp. 105–114, 254) as a local variant of a high-level blanket-bog. Small basin-bogs occur in the Huntershall area (ibid., p. 104), and extensive areas of blanket-bog are present high up in the Hope Hills.

FRESHWATER ALLUVIUM

Alluvial deposits occur along most of the modern river valleys, being particularly well developed along the River Tyne and its tributaries to the south (Figure 23). These form generally narrow strips of flood-plain alluvium and river terraces. The deposits consist of interbedded gravels, sands, silts and clays, in constantly varying proportions, though the gravels and coarse sands are common in the steeper head-waters and higher terraces and the finer deposits in the gentler, lower reaches and the flood-plain alluvium. Where favourable conditions occurred, peat deposits also form part

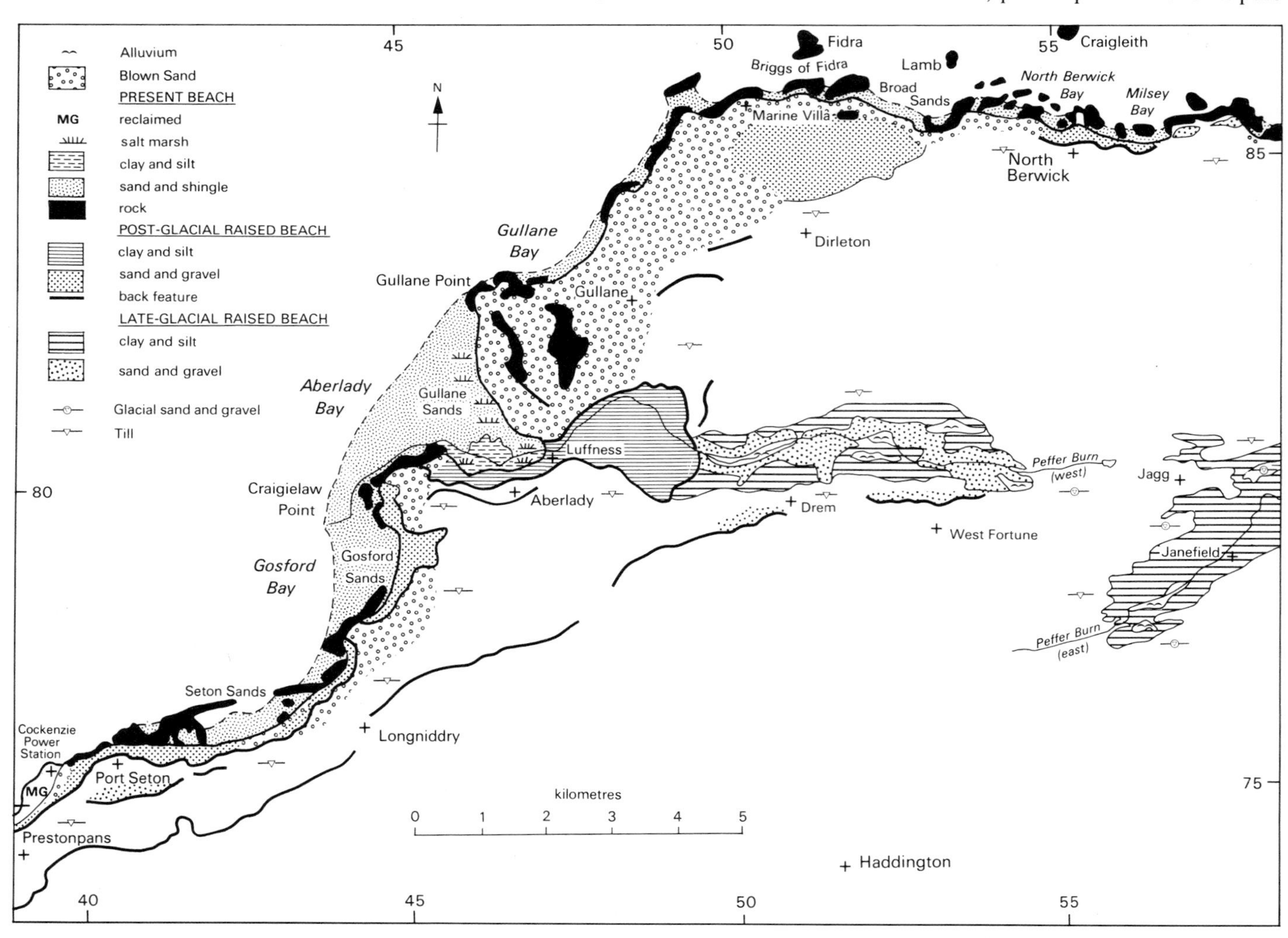

Figure 23 Map of raised beach and present shore deposits

of the alluvium. Many of the dry, glacial-drainage channels are floored by alluvial deposits. Lake deposits of laminated clay and silt, with peat, are found in flat areas infilling hollows in the glaciated surface, larger areas occurring at Dirleton and Morham. In a lake deposit in the glacial channel at Balgone (Figure 22, Locality 53), Murchison recorded deposits of peat and shell marl with *Lymnaea* associated with human skulls and bones of red deer, wild boar, aboriginal oxen and horse (Clough and others, 1910, p. 185). In many upland valleys deposits of head and hill-wash, consisting of clay, silt and stones, accompany the river alluvium.

Along three stretches of its valley, the River Tyne flows in a wide alluvial plain flanked by glacial sand and gravel deposits. These lie from Oxenfoord Mains to Winton, from Spilmersford Mains to Lennoxlove and from west of Haddington to Stevenson. The alluvium forms sloping flood-plains and dissected terraces. The present flood-plain consists mainly of fine sand and silt. The upper terraces consist of gravel and sand. Between these three areas the river flows in steep-sided valleys with narrow strips of alluvium. At the east edge of the district the River Tyne flows into a gorge cut in basalt lavas, with little or no alluvium. The gorge is a young feature cut in the rock when the original channel to the north through Beanston was blocked by till.

BLOWN SAND

Dunes of blown sand have developed in several places along the coast, generally where these are sandy post-Glacial raised beach and present beach deposits (Figure 23). Large areas of dunes occur around Gullane and Gosford. The deposits are mostly recent in age though some may date back to post-Glacial times. High dune ridges 6 to 8 m high form barriers along H.W.M. as at Gullane Bay, Gullane Sands and Broad Sands. Inland from Gullane Bay the dunes are aligned parallel to the prevailing south-west wind. The deposit is a fine-grained, cross-bedded sand, consisting mainly of quartz with about 10 per cent shell debris and 5 per cent ferromagnesian minerals. The deposits have been worked in small pits at Gullane (McAdam, 1978). ADM

CHAPTER 12

Economic geology

COAL

Coal has been mined in the district from the early part of the 13th century and possibly before then (McNeill, 1884; 1902). Apart from a few small workings in the upper part of the Calciferous Sandstone Measures, the entire output has come from the seams in the Limestone Coal Group. There are no known records of workings in the Coal Measures seams at Port Seton.

In many parts of the coalfield south-east of the Crossgatehall Fault the dip is less than 6°. This, combined with the shallowness of some of the main seams, made mining relatively easy in some areas.

At least eight seams in the Limestone Coal Group were worked in the district and up to five coals were mined in the undersea extension of the coalfield from Prestonlinks Colliery. Anderson (*in* Clough and others, 1910, pp. 188–193) gave an account of the seams being worked early in the present century together with analyses of the seams. In 1947 when the National Coal Board was formed, there were six collieries in the district. These are now all closed and the only coal being worked at present is in a large opencast operation at Blindwells, north-east of Tranent. There, several seams in the upper part of the Limestone Coal Group are being extracted, including stoops (pillars) remaining from old workings (Plate 8). In the vicinity of Tranent the Great Seam and some of the lower coals were worked over several centuries and these old shallow workings are a potential hazard to building development as their extent is unknown.

HYDROCARBONS

Gas and oil were encountered in the upper part of the Calciferous Sandstone Measures at 142.09 and 144.93 m respectively in the Spilmersford Borehole and analyses of samples are given by Davies (1974, pp. 8–10).

IRONSTONE

In the 19th century, blackband ironstone occurring in the Limestone Coal Group was mined in the district. The most important supply came from mines in the Preston–Macmerry area, south-east of Tranent. The bed which was worked, occurring between the Five Foot and Ball coals, was of excellent quality and output reached 100 tons per day but operations ceased around 1880 (Macgregor and others, 1920, pp. 179–181).

SEATCLAY

Prior to 1880, seatclays associated with several coals in the Limestone Coal Group were mined at Prestonlinks Colliery and the clay was used in the manufacture of firebricks, pipes, etc. at a fireclay works near Tranent.

CLAY

Late-Glacial clays, some of lacustrine origin, were worked in the past for the manufacture of bricks and tiles. There are records of pits near Aberlady [472 787], East Fenton [520 809], East Fortune [558 801], Gladsmuir [462 731] and East Saltoun [476 673] where clay was obtained for use in local brick and tile works.

LIMESTONE

Details, including analyses, of the limestones of economic importance in the district were given by Haldane and Simpson (1942) and Robertson and others (1949, pp. 92–96). The economically most important beds are the Upper Longcraig and the Skateraw limestones, both occurring in the lower part of the Lower Limestone Group. Both limestones were quarried extensively along the outcrop from the latter part of the 18th century into the present century but most of the quarries are now overgrown or have been filled in. The limestone was used for agricultural purposes, when ground down, and in the building industry after being burned in kilns.

Limestones in the Calciferous Sandstone Measures were also quarried, the principal ones being the Sunnyside Limestone in the North Berwick area and the Lower Longcraig Limestone near Lennoxlove.

BUILDING STONE

In the past, both sandstone and igneous rocks were extensively worked in numerous quarries in the district for the construction of local buildings. MacGregor (1945, p. 15) demonstrated the marked decline in the quarrying of sandstone during the early part of this century and the industry is now defunct.

The red and brown sandstones of the Devono-Carboniferous were extracted on a small scale at a few localities but the majority of the quarries were in the sandstones of Carboniferous age. Grey, yellow and brown stone was excavated from several horizons in the Calciferous Sandstone Measures and the Limestone Coal Group and from minor occurrences in the Lower and Upper limestone groups. Most of the quarries have been filled in or are overgrown.

Of the various igneous rocks wrought for building stone, some were quarried from the lavas and tuffs of the Garleton

Hills Volcanic Rocks (Clough and others, 1910, pp. 196–197; MacGregor, 1945, p. 30). A good stone was obtained from trachytic lavas in the Dirleton area, some of which was used in recent years for the building of the Sancta Maria Abbey, Nunraw, near Garvald.

A porphyritic trachyte quarried locally was used in the building of the older part of Haddington, and the old part of North Berwick was built of the reddish brown phonolite quarried on the south-west side of North Berwick Law.

ROADSTONE

Carboniferous igneous rocks, from intrusions and lava flows, and to a lesser extent Ordovician greywackes, have been quarried in many parts of the district for use as roadstone. The characteristics of some of the rocks so used have been described by Bailey and Anderson (*in* Clough and others, 1910, pp. 198–201).

The Ordovician greywackes, which are present south-east of the Lammermuir Fault, have been used for roadstone locally for a long time. Several quarries have been excavated in them, the oldest being those along the sides of the Roman road, Dere Street, west of Duns Law [460 575] and Turf Law [472 561] in the Lammermuir Hills. The stone was evidently used for the construction and maintenance of this ancient highway.

Of the various types of Carboniferous lavas in the district, basalt and mugearite have seldom been worked, but trachyte was quarried at the former Craigs Quarry [507 834] and trachyandesite was extracted in the Garleton Hills and is currently being worked at Bangley Quarry [487 752].

Intrusive rocks have been used in large quantities for roadstone in the district. Until recently, considerable excavation took place of the phonolite intrusion of Traprain Law [583 749] on the eastern margin of the district. An intrusive trachyte, was formerly worked at Pencraig Quarry [573 765] and more recently in the nearby Markle Mains Quarry [587 770]. The Gosford Bay teschenite sill from Gosford Quarry [440 776] produced a good quality roadstone. Fine-grained intrusions of basanite (analcime-basalt) were quarried at several localities in the Gifford area. A dolerite intrusion was worked at West Fenton [494 807]. Quartz-dolerite dykes were formerly used for roadstone from the Tranent and Longniddry areas.

SAND AND GRAVEL

Sand and gravel deposits have been worked at several localities in the district and details have been given by Haldane (1948, pp. 21–23), Goodlet (1970, pp. 69–81) and McAdam (1978). Deposits of economic importance occur in a discontinuous belt on the north-west flanks of the Lammermuir Hills from Tynehead near the western margin of the district to the area south of Gifford. Extraction is currently taking place (1984) at Long Yester [533 641] and Keith Marischal [451 641].

At lower levels farther north, glacial sands and gravels also occur in the valley of the Tyne Water between Pathhead and Ormiston and in the Pencaitland and East Fortune areas.

The deposits are partly in the form of terraces and partly occur as mounds and ridges. Near the Lammermuir Hills the predominant constituent is medium to coarse gravel, composed mainly of greywacke pebbles and boulders. Farther north, fine to medium sands occur with lenticular beds of gravel, the pebbles being mainly of sandstone and smaller proportions of greywacke, limestone and igneous rocks. Layers of coal and carbonaceous shale fragments are a widespread feature of these deposits.

Blown sand from the dunes near Gullane has been worked at times.

HEMATITE

The Garleton Hematite Mine (Flett *in* Macgregor and others, 1920, p. 212), which ceased production in 1876, is situated near the Hopetoun Monument [5007 7642] in the western part of the Garleton Hills. The country rock is trachyte lava and the vein was traced in the workings for 270 m along the strike in a north-north-west direction, proving to be irregular in thickness and reaching a maximum of 1.8 m. The ore occupied a well-defined fissure dipping to the ENE at 80° to 90° and the vein-breccia was cemented mainly by baryte. The vein thinned to uneconomic proportions at a depth of 85 m and extraction ceased. In 1874 the mine produced over 10 000 tons of good quality ore.

PEAT

The most extensive area of thick peat in the district is at Fala Moor [430 580] where a high-level, blanket bog covers about 130 ha, with a maximum known thickness of 7.9 m. Other, smaller areas occur on the high ground in the southern part of the district, including one near Huntershall [471 584] where peat has been extracted. WT

GROUNDWATER

The Calciferous Sandstone Measures are the main water-bearing strata and four wells in the northern part of the district are at present extracting water from these rocks. Two of these wells are capable of supplying up to 1000 m^3/day (240 000 gallons per day), the remainder less than 500 m^3/day. Water quality is generally good although a borehole near Aberlady produces water with a higher than normal iron-content.

Little is known about the aquifer characteristics of the remainder of the Carboniferous. The Lower Limestone Group yields a small amount of groundwater from a shallow borehole at Longniddry and at Pencaitland the Limestone Coal Group provides 650 m^3/day from a single well. Old coal workings in this group are commonly flooded and hold vast amounts of low quality water in storage. With the exception of a maltings at Haddington, all the wells provide water for irrigation. Several trial water boreholes in the volcanic rocks of the Garleton Hills produced very low yields indicating that, along with the Ordovician and Silurian greywackes and shales, these rocks are generally unsuitable for groundwater abstraction. DFB

REFERENCES

Anderson, E. M. 1951. *The dynamics of faulting and dyke formation with application to Britain* (2nd edition). (Edinburgh: Oliver and Boyd.)

Battey, M. H. and Moss, A. A. 1962. Powellite from Traprain Law, Haddington, Scotland. *Mineral. Mag.*, Vol. 33, 158–167.

Bennett, J. A. E. 1945. Some occurrences of leucite in East Lothian. *Trans. Edinburgh Geol. Soc.*, Vol. 14, 34–52.

Bennett, J. R. P. 1969. Results of geophysical surveys in the Haddington area. *Inst. Geol. Sci. Intern. Rep.*, GP/AG/70/13.

Bremner, A. 1942. The origin of the Scottish river system. *Scott. Geogr. Mag.*, Vol. 58, 15–20, 54–59, 99–103.

Burgess, I. C. 1960. Fossil soils of the Upper Old Red Sandstone of south Ayrshire. *Trans. Geol. Soc. Glasgow*, Vol. 24, 138–153.

Cameron, I. B. and Stephenson, D. 1985. The Midland Valley of Scotland (3rd edition). *Brit. Reg. Geol. Br. Geol. Surv.*

Chapman, N. A. 1976. Inclusions and megacrysts from undersaturated tuffs and basanites, East Fife, Scotland. *J. Petrol.*, Vol. 17, 472–498.

Clark, R. H. 1956. A petrological study of the Arthur's Seat Volcano. *Trans. R. Soc. Edinburgh*, Vol. 63, 37–70.

Clough, C. T., Barrow, G., Crampton, C. B., Maufe, H. B., Bailey, E. B. and Anderson, E. M. 1910. The geology of East Lothian. *Mem. Geol. Surv. G.B.*

Craig, G. Y. (Editor). 1965. *The geology of Scotland.* (1st edition). (Edinburgh: Oliver and Boyd.)

— and Walton, E. K. 1959. Sequence and structure in the Silurian rocks of Kirkcudbrightshire. *Geol. Mag.*, Vol. 96, 209–220.

— and Duff, P. M. D. (Editors). 1975. *The geology of the Lothians and south east Scotland.* (Edinburgh: Edinburgh Geological Society, Scottish Academic Press.)

Currie, E. D. 1954. Scottish Carboniferous goniatites. *Trans. R. Soc. Edinburgh*, Vol. 62, 527–602.

Davies, A. 1974. The Lower Carboniferous (Dinantian) sequence at Spilmersford, East Lothian, Scotland. *Bull. Geol. Surv. G.B.*, No. 45, 1–38.

— McAdam, A. D. and Cameron, I. B. In press. The geology of the Dunbar district. *Mem. Br. Geol. Surv.*

Day, T. C. 1914. Notes on the Hummell Rocks, Gullane. *Trans. Edinburgh Geol. Soc.*, Vol. 10, 114–119.

— 1916a. The Cheese Bay Sill, Gullane. *Trans. Edinburgh Geol. Soc.*, Vol. 10, 249–260.

— 1916b. The breccias of Cheese Bay and the yellow conglomerates of Weaklaw. *Trans. Edinburgh Geol. Soc.*, Vol. 10, 261–275.

— 1923. A new volcanic vent and other new geological features on the shore, Weaklaw, near Gullane. *Trans. Edinburgh Geol. Soc.*, Vol. 11, 185–192.

— 1925a. The Leithies, North Berwick, a small laccolite with unusual intrusive phenomena. *Trans. Edinburgh Geol. Soc.*, Vol. 11, 300–307.

— 1925b. Two unrecorded volcanic vents on the shore east of North Berwick. *Trans. Edinburgh Geol. Soc.*, Vol. 11, 338–345.

— 1928a. The volcanic vents on the shore between North Berwick and Tantallon Castle. *Trans. Edinburgh Geol. Soc.*, Vol. 12, 41–52.

— 1928b. Metasomatism in basalt, near Eel Burn, North Berwick. *Trans. Edinburgh Geol. Soc.*, Vol. 12, 117–121.

— 1930a. An igneous dyke in the quartz-banakite of Bangly Quarry near Haddington. *Trans. Edinburgh Geol. Soc.*, Vol. 12, 256–259.

— 1930b. Chemical analyses of thirteen igneous rocks of East Lothian. *Trans. Edinburgh Geol. Soc.*, Vol. 12, 263–266.

— 1932a. The teschenite of Point Garry, North Berwick. *Trans. Edinburgh Geol. Soc.*, Vol. 12, 334–337.

— 1932b. Large sandstone xenolith within the phonolite of Traprain Law. *Trans. Edinburgh Geol. Soc.*, Vol. 12, 338–341.

— 1932c. Volcanic vents at Longskelly Rocks and Yellow Craig Plantation, west of North Berwick. *Trans. Edinburgh Geol. Soc.*, Vol. 12, 376–381.

De Souza, H. A. F. 1974. Potassium-argon ages of Carboniferous igneous rocks from East Lothian and the south of Scotland. Unpublished MSc thesis, University of Leeds.

— 1979. The geochronology of Scottish Carboniferous volcanism. Unpublished PhD thesis, University of Edinburgh.

Donner, J. J. 1963. The Late- and Post-Glacial raised beaches in Scotland, 2. *Ann. Acad. Sci. Fenn. Sar. A3 Geol. Geogr.*, Vol. 68, 1–13.

Flint, R. F. 1957. *Glacial and Pleistocene geology.* (New York: Wiley.)

Floyd, J. D. 1982. Stratigraphy of a flysch succession: the Ordovician of W Nithsdale, SW Scotland. *Trans. R. Soc. Edinburgh Earth Sci., Vol. 73, 1–9.*

Forsyth, I. H. and Chisholm, J. I. 1977. The geology of East Fife. *Mem. Geol. Surv. G.B.*

Francis, E. H. 1960. Intrusive tuffs related to the Firth of Forth volcanoes. *Trans. Edinburgh Geol. Soc.*, Vol. 18, 32–50.

George, T. N. 1960. The stratigraphical evolution of the Midland Valley. *Trans. Geol. Soc. Glasgow*, Vol. 24, 32–107.

— Johnson, G. A. L., Mitchell, M., Prentice, J. E., Ramsbottom, W. H. C., Sevastopulo, G. D. and Wilson, R. B. 1976. A correlation of Dinantian rocks in the British Isles. *Spec. Rep. Geol. Soc. London*, No. 7.

Goodlet, G. A. 1957. Lithological variation in the Lower Limestone Group in the Midland Valley of Scotland. *Bull. Geol. Surv. G.B.*, No. 12, 52–65.

— 1959. Mid-Carboniferous sedimentation in the Midland Valley of Scotland. *Trans. Edinburgh Geol. Soc.*, Vol. 17, 217–240.

— 1970. Sands and gravels of the southern counties of Scotland. *Rep. Inst. Geol. Sci.*, No. 70/4.

Graham, A. M. and Upton, B. G. J. 1978. Gneisses in diatremes, Scottish Midland Valley: petrology and tectonic implications. *J. Geol. Soc. London*, Vol. 135, 219–226.

Greig, D. C. 1971. The south of Scotland (3rd edition). *Br. Reg. Geol. Inst. Geol. Sci.*

Haldane, D. 1948. Sands and gravels of Scotland. Quarter-inch Sheet 15, Fife–the Lothians–Berwickshire. *Wartime Pamphlet Geol. Surv. G.B.*, No. 30, Part 4.

— and Simpson, J. B. 1942. Limestones of Scotland: East-Central Scotland. *Wartime Pamphlet Geol. Surv. G.B.*, No. 13, Area 3.

Hatch, F. H. 1892. *Text-book of petrology.* (London: Sonnenschein.)

HESSELBO, S. P. and TREWIN, N. H. 1984. Deposition, diagenesis and structures of the Cheese Bay Shrimp Bed, Lower Carboniferous, East Lothian. *Scott. J. Geol.*, Vol. 20, 281–296.

HOWELL, H. H., GEIKIE, A. and YOUNG, J. 1866. The geology of East Lothian. *Mem. Geol. Surv. G.B.*

HOWELLS, M. F. 1969. Cryptovents and allied structures in Carboniferous strata between Port Seton and Aberlady, East Lothian. *Scott. J. Geol.*, Vol. 5, 1–10.

KELLING, G. 1961. The stratigraphy and structure of the Ordovician rocks of the Rhinns of Galloway. *Q.J. Geol. Soc. London,* Vol. 117, 37–75.

KENDALL, P. F. and BAILEY, E. B. 1908. The glaciation of East Lothian south of the Garleton Hills. *Trans. R. Soc. Edinburgh,* Vol. 46, 1–31.

KENNEDY, W. Q. 1958. The tectonic evolution of the Midland Valley of Scotland. *Trans. Geol. Soc. Glasgow,* Vol. 23, 106–133.

KIRBY, R. P. 1968. The ground moraines of Midlothian and East Lothian. *Scott. J. Geol.*, Vol. 4, 209–220.

LAGIOS, E. 1984. A geophysical study of the East Lothian volcanics, southeast Scotland. *Earth Planet. Sci. Lett.*, Vol. 67, 205–210.

— and HIPKIN, R. G. 1982. A geophysical approach to the granite batholith under the eastern Southern Uplands. *Pure Appl. Geophys.*, Vol. 120, 375–388.

LAMONT, A. and LINDSTROM, M. 1957. Arenigian and Llandeilian cherts identified in the Southern Uplands of Scotland by means of conodonts etc. *Trans. Edinburgh Geol. Soc.*, Vol. 17, 60–70.

LAPWORTH, C. 1889. On the Ballantrae rocks of south Scotland and their place in the Upland Sequence. *Geol. Mag.*, Dec. 3, Vol. 6, 59–69.

LEEDER, M. 1973. Sedimentology and palaeogeography of the Upper Old Red Sandstone in the Scottish Border Basin. *Scott. J. Geol.*, Vol. 9, 117–144.

LINTON, D. L. 1951. Problems of Scottish scenery. *Scott. Geogr. Mag.*, Vol. 67, 65–85.

McADAM, A. D. 1974. The petrography of the igneous rocks in the Lower Carboniferous (Dinantian) at Spilmersford, East Lothian, Scotland. *Bull. Geol. Surv. G.B.*, No. 45, 39–46.

— 1978. Sand and gravel resources of the Lothian Region of Scotland. *Rep. Inst. Geol. Sci.*, No. 78/1.

— 1983. Preliminary report on the sand and gravel deposits of the Humbie–East Linton area, Lothian Region. *Inst. Geol. Sci. Intern. Rep.* SL 83/4.

MACDONALD, R. 1975. Petrochemistry of the early Carboniferous (Dinantian) lavas of Scotland. *Scott. J. Geol.*, Vol. 11, 269–314.

— GOTTFRIED, D., FARRINGTON, M. J., BROWN, F. W. and SKINNER, N. G. 1981. Geochemistry of a continental tholeiite suite: late Palaeozoic quartz-dolerite dykes of Scotland. *Trans. R. Soc. Edinburgh Earth Sci.*, Vol. 72, 57–74.

MACGREGOR, A. G. 1928. The classification of Scottish Carboniferous olivine-basalts and mugearites. *Trans. Geol. Soc. Glasgow,* Vol. 18, 324–360.

— 1939. The term 'Plagiophyre'. *Bull. Geol. Surv. G.B.*, No. 1, 99–103.

— 1945. The mineral resources of the Lothians. *Wartime Pamphlet Geol. Surv. G.B.*, No. 45.

— 1966. Divisions of the Carboniferous on Geological Survey Scottish maps. *Bull. Geol. Surv. G.B.*, No. 16, 127–130.

— and ENNOS, F. R. 1922. The Traprain Law Phonolite. *Geol. Mag.*, Vol. 59, 514–523.

MACGREGOR, M., LEE, G. W. and WILSON, G. V. 1920. The iron ores of Scotland. *Mem. Geol. Surv. G.B., Min. Resour.*, Vol. 11.

— and MACGREGOR, A. G. 1948. The Midland Valley of Scotland (2nd edition). *Br. Reg. Geol. Geol. Surv. G.B.*

McKERROW, W. G., LEGGETT, J. K. and EALES, M. H. 1977. Imbricate thrust model of the Southern Uplands of Scotland. *Nature, London,* Vol. 267, 237–239.

McNEILL, P. 1884. *Tranent and its surroundings* (2nd edition). (Tranent: McNeill.)

— 1902. *Prestonpans and vicinity.* (Tranent: McNeill.)

MARTIN, N. R. 1955. Lower Carboniferous volcanism near North Berwick. *Bull. Geol. Surv. G.B.*, No. 7, 90–100.

MITCHELL, G. H., WALTON, E. K. and GRANT, D. (Editors). 1960. *Edinburgh Geology, an excursion guide.* (Edinburgh and London: Oliver and Boyd.)

MITCHELL, G. H. and MYKURA, W. 1962. The geology of the neighbourhood of Edinburgh (3rd edition). *Mem. Geol. Surv. G.B.*

MUIR, I. D. and TILLEY, C. E. 1961. Mugearites and their place in the alkali igneous rock series. *J. Geol.*, Vol. 69, 186–203.

NEVES, R. and IOANNIDES, N. 1974. Palynology of the Lower Carboniferous (Dinantian) of the Spilmersford, Borehole, East Lothian, Scotland. *Bull. Geol. Surv. G.B.*, No. 45, 73–97.

— GUEINN, K. J., CLAYTON, G., IOANNIDES, N. S., NEVILLE, R. S. W. and KRUSZEWSKA, K. 1973. Palynological correlations within the Lower Carboniferous of Scotland and Northern England. *Trans. R. Soc. Edinburgh,* Vol. 69, 23–70.

PEACH, B. N. 1908. Monograph on the higher crustacea of the Carboniferous rocks of Scotland. *Pal. Mem. Geol. Surv. G.B.*

— and HORNE, J. 1899. The Silurian rocks of Britain. Vol. 1. Scotland. *Mem. Geol. Surv. G.B.*

RAGG, J. M. and FUTTY, D. W. 1967. The soils of the country around Haddington and Eyemouth. *Mem. Soil Surv. G.B.*

RAMSBOTTOM, W. H. C. 1977. Correlation of the Scottish Upper Limestone Group (Namurian) with that of the North of England. *Scott. J. Geol.*, Vol. 13, 327–330.

— CALVER, M. A., EAGER, R. M. C., HODSON, F., HOLLIDAY, D. W., STUBBLEFIELD, C. J. *and* WILSON, R. B. 1978. A correlation of Silesian rocks in the British Isles. *Spec. Rep. Geol. Soc. London,* No. 10.

RICHEY, J. E., ANDERSON, E. M. *and* MACGREGOR, A. G. 1930. The geology of North Ayrshire. *Mem. Geol. Surv. G.B.*

ROBERTSON, T., SIMPSON, J. B. and ANDERSON, J. G. C. 1949. The limestones of Scotland. *Mem. Geol. Surv. G.B., Mineral Resources,* Vol. 35.

SIMPSON, J. B. 1928. Notes on the geology of the Kidlaw district, East Lothian. *Trans. Edinburgh Geol. Soc.*, Vol. 12, 111–113.

SISSONS, J. B. 1958. The deglaciation of part of East Lothian. *Trans. Inst. Br. Geogr.*, Vol. 25, 59–77.

— 1967. *The evolution of Scotland's scenery.* (Edinburgh: Oliver and Boyd.)

— 1974. The Quaternary in Scotland: a review. *Scott. J. Geol.*, Vol. 10, 311–337.

SMITH, S. M. 1972. Palaeoecology of post-glacial beaches in East Lothian. *Scott. J. Geol.*, Vol. 8, 31–49.

SMITH, T. E. 1967. A preliminary study of sandstone sedimentation in the Lower Carboniferous of the Tweed Basin. *Scott. J. Geol.*, Vol. 3, 282–305.

SNODGRASS, C. P. *The county of East Lothian. The Third Statistical Account of Scotland.* (Edinburgh: Oliver and Boyd.)

SUTHERLAND, D. S. (Editor). 1982. *The igneous rocks of the British Isles.* (Chichester: Wiley.)

TILLEY, C. E. and MUIR, I. D. 1964. Intermediate members of the oceanic basalt-trachyte association. *Geol. Foren Stock. Forhand.*, Vol. 85, 436–444.

Tomkeieff, S. I. 1937. Petrochemistry of the Scottish Carboniferous–Permian igneous rocks. *Bull. Volcan.*, (2), Vol. 1, 59–87.

— 1952. Analcite-trachybasalt inclusions in the phonolite of Traprain Law. *Trans. Edinburgh Geol. Soc.*, Vol. 15, 360–373.

Traquair, R. H. 1907. Report on fossil fishes collected by the Geological Survey of Scotland from shales exposed on the shore near Gullane, East Lothian. *Trans. R. Soc. Edinburgh,* Vol. 46, 103–117.

Tulloch, W. and Walton, H. S. 1958. The geology of the Midlothian Coalfield. *Mem. Geol. Surv. G.B.*

Upton, B. G. J. 1969. *Field excursion guide to the Carboniferous volcanic rocks of the Midlothian Valley of Scotland.* (Oxford: International Association of Volcanology and Chemistry.)

— Aspen, P and Chapman, N. A. 1983. The upper mantle and deep crust beneath the British Isles: evidence from inclusions in volcanic rocks. *J. Geol. Soc. London,* Vol. 140, 105–122.

Walton, E. K. 1955. Silurian greywackes in Peeblesshire. *Proc. R. Soc. Edinburgh,* B65, 327–357.

— 1961. Some aspects of the succession and structure in the Lower Palaeozoic rocks of the Southern Uplands of Scotland. *Geol. Rundsch.,* Vol. 50, 63–77.

Wilson, R. B. 1967. A study of some Namurian marine faunas of Central Scotland. *Trans. R. Soc. Edinburgh,* Vol. 66, 445–490.

— 1974. A study of the Dinantian marine faunas of south-east Scotland. *Bull. Geol. Surv. G.B.,* No. 46, 35–65.

Young, B. R. 1903. An analcite diabase and other rocks from Gullane. *Trans. Edinburgh Geol. Soc.,* Vol. 8, 326–335.

INDEX

HER MAJESTY'S STATIONERY OFFICE

HMSO publications are available from:

HMSO Publications Centre
(Mail and telephone orders)
PO Box 276, London SW8 5DT
Telephone orders (01) 622 3316
General enquiries (01) 211 5656

HMSO Bookshops
49 High Holborn, London WC1V 6HB
(01) 211 5656 (Counter service only)
258 Broad Street, Birmingham B1 2HE (021) 643 3757
Southey House, 33 Wine Street, Bristol BS1 2BQ
(0272) 24306/24307
9 Princess Street, Manchester M60 8AS (061) 834 7201
80 Chichester Street, Belfast BT1 4JY (0232) 234488
13a Castle Street, Edinburgh EH2 3AR (031) 225 6333

HMSO's Accredited Agents
(see Yellow Pages)

And through good booksellers

BRITISH GEOLOGICAL SURVEY

Keyworth, Nottinghamshire NG12 5GG

Murchison House, West Mains Road, Edinburgh EH9 3LA

The full range of Survey publications is available through the Sales Desks at Keyworth and Murchison House. Selected items are stocked by the Geological Museum Bookshop, Exhibition Road, London SW7 2DE; all other items may be obtained through the BGS Information Point in the Geological Museum. All the books are listed in HMSO's Sectional List 45. Maps are listed in the BGS Map Catalogue and Ordnance Survey's Trade Catalogue. They can be bought from Ordnance Survey Agents as well as from BGS.

On 1 January 1984 the Institute of Geological Sciences was renamed the British Geological Survey. It continues to carry out the geological survey of Great Britain and Northern Ireland (the latter as an agency service for the government of Northern Ireland), and of the surrounding continental shelf, as well as its basic research projects. It also undertakes programmes of British technical aid in geology in developing countries as arranged by the Overseas Development Administration.

The British Geological Survey is a component body of the Natural Environment Research Council.

Printed for Her Majesty's Stationery Office by Linneys Colour Print Ltd.
Dd 738605 C20 1/86 46008